"This is a much-needed book that will be of use to those tasked with the care of our precious stock of pre-1919 brick-built properties by helping them to recognise and better understand the principal causes of decay and failure of bricks, brickwork, terracotta and various traditional mortars utilised in their construction and articulation. Rather than exploring techniques of repair and maintenance, this work deliberately sets out to concentrate on the many inter-connected historical, technical and scientific aspects, studied in a logical and comprehensive manner, specifically intended to provide a more informed and methodical approach to surveying historic brickwork."

– Dr Gerard Lynch, freelance masonry specialist, educator and author, United Kingdom

T0187687

Survey and Assessment of Traditionally Constructed Brickwork

The principal aim of this book is to inform the practice of surveying traditionally constructed brickwork. It aims to ensure that those who have a cause to survey buildings constructed of traditional (pre-1919) brickwork have a well-developed, clear methodology for undertaking effective, comprehensive surveys. The book will help readers gain the proper level of knowledge, expertise and skill to be able to survey these buildings correctly; to recognise not just defects within the brickwork itself but also, crucially, the cause of these defects. Experienced author Moses Jenkins presents a clear and concise methodology for the survey of brickwork in existing buildings with coverage including:

- identifying repair needs
- understanding the cause of decay within brickwork
- ascertaining the age and significance of traditionally constructed brickwork, and
- identifying significant technical and decorative features.

Despite the extensive use of traditionally constructed brick worldwide, the knowledge and expertise to properly survey buildings of this type has not, to date, been laid out in a methodical way. This book addresses this deficiency and provides an invaluable resource to all those who survey and assess buildings constructed of brickwork. This includes building professionals such as architects, surveyors, engineers and those working in building conservation, but also construction managers and students studying built environment disciplines.

Moses Jenkins is a Senior Technical Officer with Historic Environment Scotland and has led its technical research into bricks and brickwork since 2008. He received his PhD from the University of Dundee in 2016 and has published numerous articles and four previous books on brickwork, surveying historic buildings and conservation.

Survey and Assessment of Traditionally Constructed Brickwork

Moses Jenkins

Routledge
Taylor & Francis Group

LONDON AND NEW YORK

Cover image: © Moses Jenkins

First published 2022
by Routledge
2 Park Square, Milton Park, Abingdon, Oxon OX14 4RN

and by Routledge
605 Third Avenue, New York, NY 10158

Routledge is an imprint of the Taylor & Francis Group, an informa business

© 2022 Moses Jenkins

The right of Moses Jenkins to be identified as author of this work has been
asserted by him in accordance with sections 77 and 78 of the Copyright,
Designs and Patents Act 1988.

British Library Cataloguing-in-Publication Data
A catalogue record for this book is available from the British Library

Library of Congress Cataloging-in-Publication Data
Names: Jenkins, Moses, author.
Title: Surveying traditionally constructed brickwork/Moses Jenkins.
Description: Abingdon, Oxon; New York, NY: Routledge, 2022. |
Includes bibliographical references and index.
Identifiers: LCCN 2021046840 (print) | LCCN 2021046841 (ebook) |
Subjects: LCSH: Bricks–Inspection. | Masonry–Inspection. |
Building, Brick. | Building inspection.
Classification: LCC TA679 .J46 2022 (print) |
LCC TA679 (ebook) | DDC 693/.21–dc23/eng/20211117
LC record available at https://lccn.loc.gov/2021046840
LC ebook record available at https://lccn.loc.gov/2021046841

ISBN: 978-0-367-55582-5 (hbk)
ISBN: 978-0-367-55583-2 (pbk)
ISBN: 978-1-003-09416-6 (ebk)

DOI: 10.1201/9781003094166

Typeset in Bembo
by Newgen Publishing UK

Contents

Acknowledgements

With any substantive piece of work, the support of others is a necessity. The generosity of those who have helped with this book, however, has exceeded all expectations. Many have given freely of their experience, knowledge and time, without which this book would not be what it is today. Any errors and omissions, which will certainly be present, are all my full responsibility.

Several friends and colleagues helped by reading sections of this work. My thanks go to:

> Dr Gerard Lynch for reading the section on gauged brickwork, and for many years of sharing knowledge and expertise
> Bryan Hindle for reading the section on cavity walls
> Adam Frost for reading the section on digital documentation
> Eamon Gilson for reading the section on BIM
> Steve Wood for reading the section on cracks and structural movement
> Dr Maureen Young for reading the section on thermography
> Dr Callum Graham for reading the section on mortar analysis.

Material to use as the basis of several case studies came from friends and colleagues around the world:

> Bill Revie of CMC provided the material that formed the basis of Case study 8.
> Roz Artis provided the material which formed the basis of Case study 7.
> Nicola Ashurst provided the material that formed the basis of Case study 1.
> Graham Hickey provided the material which formed the basis of Case study 2.
> Sun Zheng and Jiangtao Xie kindly allowed me to summarise their work for Case study 9.

Thanks to those who helped out with photos, including those mentioned above, Jason Turrall at Furness Brick, everyone at Georgian Brickwork, Jill Fairweather, Andrew Mulhern at the Northern Lighthouse Board, Mathew Slocomb and the SPAB, David Pickles and Historic England.

Thanks to everyone at Routledge for accepting the proposal and all the work on the book, especially Patrick Hetherington for putting up with all the delays.

Thanks to all my family for the support over the years.

Most of all, thank you to my beloved wife – this book would not have happened without you.

Introduction

A significant proportion of the buildings extant in the United Kingdom today are of traditional construction. As an average across the United Kingdom, in the region of 20 per cent of all buildings were built before 1919. This amounts to almost five million structures, a large proportion of which are constructed partly of brick. Despite this extensive use of traditionally constructed brick, the knowledge and expertise to survey and assess buildings of this type has not, to date, been laid out in a concise, methodical way. Brickwork has been used in the construction of some of the finest historic buildings in the United Kingdom: Hampton Court Palace, St Pancras Railway Station, Templeton's Carpet Factory and many more buildings of great architectural and historic significance. Without the proper level of knowledge, expertise and skills to survey these buildings correctly, to recognise not just defects within the brickwork itself, but also the causes of these defects, buildings constructed of traditional brickwork will not receive the repair and maintenance they require to continue to be of value into the future. The principal aim of this book is to provide a foundation of knowledge to inform the surveying and recording of traditionally constructed brickwork. The book aims to ensure that those who have a cause to survey buildings constructed of brickwork have a well-developed, clear methodology for undertaking effective, comprehensive surveys.

Any work significant in scope must begin with definitions. It is definitions that set the parameters within which the content will be constrained. For the purposes of this book, several terms must be defined at the outset. Clearly it is necessary to define what is meant by the term "brick". For the purposes of this book, a brick is defined as a unit of fired clay, generally a rectangular prism in shape and able to be held in one hand, which is used in the construction of a variety of structures. It will be immediately apparent that defining even the humble brick presents challenges: as will be seen, not all bricks are rectangular prisms. There are also numerous subdivisions that can be applied to bricks, some based on the method used in their manufacture (such as pressed facing), others based on their use (such as engineering bricks). For the purpose of the subsequent discussion, the above will serve as the definition of a brick in the broadest sense.

However, bricks alone do not a structure make. To bed bricks, mortar is required. Mortar may be defined as a mixture consisting of a binder, water and

DOI: 10.1201/9781003094166-1

aggregate that hardens, or sets, following a chemical reaction. This brings into consideration the third definition required at the outset of this work, that of brickwork. "Brickwork" is the term used to describe bricks and mortar when they are used together in construction.

It will be apparent from the title of this book that it is not brickwork of all ages that is considered here, but rather only that which may be defined as being traditionally constructed. The definition of traditional construction is widely debated, with some using a particular period of time to delineate traditional from modern practice and others using craft practice for this purpose. While both have merit, in this book the term "traditionally constructed" (often shortened to simply "traditional") will be used to refer to brickwork constructed before the year 1919. This is somewhat arbitrary, as non-traditional construction methods, including early cavity walls and cement-based mortars, were used in the period pre-1919, with traditional craft practices continuing in use in some cases until the present day. As 1919 is the cut-off date used by a number of heritage bodies, this is the definition used in the proceeding work.

The final aspect of the title of this book that requires definition at the outset is the action of survey. Survey may be defined as a detailed evaluation of the construction and condition of a building or structure. However, for the purpose of this work, a wider definition of the term is required. Despite being of considerable use to those undertaking survey in the professional sense described above, there are many professions, occupations, trades and interested parties who may also have a need to assess, interpret and record aspects of traditionally constructed brickwork. The definition of surveying for the purpose of this work, therefore, is any inspection, assessment or consideration of traditionally constructed brickwork.

As this book may be utilised by a variety of building professionals, including architects, surveyors, construction managers and those studying and learning such professions, no discussion is included with regard to the specific legal and professional requirements they may be required to meet. For example, standard construction contracts or the legal responsibilities of a Chartered Surveyor are not discussed. Several excellent works on these subjects exist already and are referred to in the References and Further Reading section at the end of this chapter. Likewise, many of the generalities of surveying, recording and assessing structures are not discussed, except in respect of the specifics of traditionally constructed brickwork. Again, many excellent works provide this knowledge and are referred to throughout.

For those who are inexperienced with the survey or assessment of traditional brickwork, this book provides a basis to conduct such work. Those who have a relatively low starting point in terms of knowledge of traditional brickwork will therefore benefit from studying this book sequentially. For professionals who have experience of traditional construction and traditional brickwork in particular, the book is set out in such a way that enables relevant sections to be consulted as required.

1 Survey and assessment of traditional brickwork

Prior to any survey and assessment work taking place, it is vital that the aims and objectives of the work are set out. It is also important that a clear methodology for assessing and surveying the brickwork is undertaken. This chapter sets out some of these preliminary considerations and attempts to provide a loose framework whereby traditional brickwork can be assessed thoroughly. Having undertaken survey work, it is also important to consider how it will be presented. Lastly, considerations around access to brickwork are important and are likewise considered here. The methodology set out in this chapter is not meant to be rigid but, like brickwork itself, it is intended to have a degree of flexibility in approach.

1.1 Establishing survey aims, objectives and purpose

Prior to conducting any survey or assessment of a traditionally constructed brick structure, it is necessary to establish what the aims, objectives and purpose of the work are. The level of detail which a survey of traditional brickwork will entail is often dependent upon the purpose for which the results of the survey will be employed. A simple condition survey to ascertain the broad physical condition of a building will not necessarily need to employ the more specialised analytical techniques described in Chapter 7. Alternatively, a comprehensive assessment of both the condition and heritage significance of a building may require a detailed knowledge of technical and aesthetic aspects of traditional brickwork with attendant use of analytical techniques and complex recording. It is for the professional who is undertaking the survey to decide in conjunction with those commissioning the work what level of detail and specialist analytics are required when conducting a survey. This will often be determined by the purpose of the survey and assessment, and the use to which the results will be put. The following subsections discuss some broad considerations around different survey types.

1.1.1 Survey as part of maintenance

One of the cornerstones of planned maintenance of a structure is regular inspection. Buildings that use brickwork as part of their construction are no

DOI: 10.1201/9781003094166-2

exception to this, and survey of brickwork as part of regular maintenance is one of the most common occasions where brickwork is assessed. It is generally advisable to carry out a thorough and structured inspection of brickwork at least once a year, with other parts of a building such as gutters and roofs requiring assessment at least every six months. Further guidance on inspection and maintenance can be found in the References and Further Reading section at the end of this chapter, in particular Watt (2007).

Regular maintenance surveys are likely to focus on the second stage of the three-stage methodology set out below. An overall understanding of brickwork as set out in stage 1 will be important, but in a building that is regularly inspected this knowledge is likely to already have been gathered and recorded. For example, if survey is part of planned maintenance, details such as brick type, bond, gauge and so on will not have changed. The defects and decay of bricks and brickwork discussed in Chapter 6 are likely to arise, in part, as a result of a lack of maintenance. Inspection and survey of brickwork form a key part of any planned maintenance regime, something that is in turn critical to the long-term condition of traditional buildings. The purpose of building maintenance inspection is to identify early signs of defects, which can then be repaired, and to carry out small tasks that can prevent future deterioration. Good building maintenance is carried out on a planned and regular basis, and considers factors that can lead to the deterioration of materials or elements forming a building. Proper maintenance does require investment, but it is certainly worthwhile in avoiding larger repairs in the future. The cost of a yearly gutter and downpipe clean, for example, will be much less than the cost of scaffolding large areas to allow for the replacement of decayed bricks. Section 6.12 will show that a maintenance inspection that allows for biological growth to be removed at an early stage is likely to result in the avoidance of potentially serious damage to

Fig. 1.1 Survey that is taking place as part of planned maintenance should consider technical characteristics of the brickwork and any defects that may be present, such as the spalled bricks and ivy growth seen here.

Fig. 1.2 In this building, an internal wall has become exposed externally; there is a high likelihood that these walls are thinner than and built of poorer quality bricks, something that should be noted during a survey as part of maintenance.

brickwork in the future. The principal defects that can affect traditional brickwork are discussed in section 6.5. Particular attention should be focused on any developing sources of moisture saturation, such as failures in rainwater disposal systems and any damaging interventions that may have occurred, such as the introduction of cement pointing. A maintenance survey may highlight the need for specialist investigation as discussed in stage 3 of the methodology laid out below, but it is likely to be stage 2 that is most relevant.

Fig. 1.3 Rainwater goods and surface drainage are critical to the condition of brick-work, and survey as part of maintenance should consider the functioning of these elements.

1.1.2 Survey ahead of repair

Where signs of decay or damage are noted during survey, consideration will then need to be given to repair. The decision to repair brickwork is one that should be taken with care, and repairs only carried out which are necessary and proportionate to the problem. There is always a danger of repairs that are not planned correctly or that are in fact unnecessary resulting in greater damage to traditional brickwork in the long term. This can be seen, for example, where small, isolated patches of spalling are taken as indicative of a need to replace much larger areas of brick. This results in unnecessary loss of original fabric and increased costs. The same can be seen when the repointing of an entire elevation is proposed when, in fact, only a small area of brickwork needs to be repaired. Any survey that highlights a need for intervention should take a conservation-based approach to repair, as discussed in section 1.5 and the case study at the end of this chapter. It is also important, however, to ensure that defects are highlighted in survey work in order to allow for repairs to be carried out. Taking a conservation-based approach does not mean doing nothing; it is simply about making proportionate decisions regarding the repair needs of a traditionally constructed brick structure.

Survey and assessment of brickwork ahead of repairs taking place will require careful consideration of all aspects of the brickwork. This is likely to involve all three stages described below to allow for a thorough understanding of the bricks and brickwork prior to interventions taking place. Where bricks are to be removed from a section of brickwork and replaced, for example, this will require careful and detailed survey prior to the work taking place. Firstly, the condition of the bricks will need to be assessed. The extent of decay will have to be considered and a decision made as to which bricks are to be removed and which can remain in the wall. This decision requires a delicate balance of different, sometimes conflicting, factors. Heritage significance will need to be considered. If masonry is constructed of bricks that are of particular rarity, significance or age, the decision to remove them becomes more significant. There is rarely, if ever, a requirement to remove adjacent bricks that have not suffered decay and repair should be restricted only to those bricks that have failed. A thorough understanding of the age and significance of brickwork as discussed in Chapter 5 is crucial to informing decisions such as this, as is a broader understanding of heritage significance. Understanding the type of brick used is also of vital importance, as different brick types react in different ways to decay. If a piece of masonry has particular aesthetic or technical features, this too will inform the decision about whether to remove or retain.

There are also technical considerations that will need to be brought into the equation: is the brick still fulfilling its structural purpose? If not, it will almost certainly require removal. There is also a need to consider whether a damaged or decayed brick will allow excess moisture into surrounding brickwork or, where walls are thin, whether that decayed or damaged brick could allow moisture to penetrate into the interior of the building itself. Where this is the case, removal of the brick is likely to be the best course of action. Aesthetic considerations will also play a part in this decision, especially where no exact match can be found for the bricks to be removed. Brick matching for replacement involves a complex series of decisions, and further guidance can be found in Jenkins (2014). It will be seen from the short discussion above that many factors need to be considered when assessing brickwork where removal of bricks is being considered.

A further common repair need identified in survey is repointing. This too needs careful consideration; mortar will fail at different rates over a building depending on exposure and it is unlikely that an entire building will require repointing at the same time. As a rough guide, unless water is penetrating through a joint, it will not require repointing until it has eroded back to a greater depth than its height – that is, if the joint is 10 mm in height and it has only eroded 5 mm, then repointing may not be required. An exception to this rule would be if it was felt that inappropriate cement repointing was leading to excessive decay in adjacent bricks, in which case it may be felt that the removal of this and repointing with a mortar that uses lime as its binder would be required. When surveying ahead of repointing work, an understanding of the heritage significance of mortar and mortar joints is required. This is discussed in greater detail in Chapter 2 and section 3.4, but will include technical considerations such as the binder and aggregate used and aesthetic considerations such as the colour of

Fig. 1.4 This 18th century structure is clearly in need of repointing, but the bricks themselves are largely in good condition; a conservation approach should always be taken when surveying traditional brickwork.

Fig. 1.5 The bricks that have been removed from this building were noted during a comprehensive survey of the structure as suffering from significant decay.

the mortar and any joint finishes that may be present. This may require specialist investigation in the form of mortar analysis, as discussed in section 7.1.

A careful assessment of the work required to a structure ahead of any repair work requires a thorough understanding of the brickwork and the wider building. This includes understanding the age and significance of the brickwork, the technical features of the brick, and mortar, the materials used, the condition of the brickwork and the causes of any defects. This will require careful and thorough application of all stages of the survey described below.

1.1.3 Heritage survey

As well as survey to assess condition and repair needs, specific survey and assessment may be carried out to ascertain the heritage significance or historical importance of a brick building or a section of brickwork. The case study at the conclusion of Chapter 2 is an excellent example of traditional brickwork being surveyed in order to ascertain the prevalence and extent of survival of a given craft practice – in this case, a particular pointing technique. The survey work of Scottish brick structures that underpinned the Historic Environment Scotland guidance on brickwork is a further example of this. Where a survey is concerned primarily with survival of craft practices or heritage significance,

Fig. 1.6 Heritage significance is judged on many levels. These warehouses are the last remaining vestige of a once-thriving industry. For this reason, the heritage significance is high, together with their architectural and aesthetic qualities.

Fig. 1.7 a and b Where survey work is being undertaken to assess heritage, this may
be simply to record a structure under threat of demolition. Survey of this tall
chimney recorded its form, bond, use of special bricks and other characteristics.
Soon after, the chimney was sadly demolished.

the sections in this book regarding aesthetic and technical features and chrono-
logical development of brickwork will be of particular importance. This is not
to say that condition would not be considered, but rather that the focus is likely
to be on the first stage of the methodology laid out below.

Fig. 1.7b

Brickwork may also be surveyed in a heritage setting as part of a heritage impact assessment. This is a process undertaken to assess the impact an action will have on the heritage of a structure. It is part of the planning process, which is often carried out as part of an initial design stage. At its heart is a thorough understanding of the heritage significance of a building. In the context of traditional brickwork, the information set out in respect to both technical and aesthetic features will provide a useful basis for informing a heritage impact assessment on a traditionally constructed building. A number of resources are in place to assist those carrying out heritage impact assessments.

A heritage impact assessment is often required during applications for listed building consent or conservation area consent. A thorough understanding of the requirements of heritage legislation in the jurisdiction being worked in is vital to determining whether or not this is required. Understanding which parts of a brick structure are significant is clearly a key underpinning requirement for survey of heritage. The example shown in Figure 5.2 is a good case in point. Only a small area of original pointing survives. Any heritage impact assessment for this structure would therefore focus on the impact of proposed work on these surviving areas of pointing.

1.1.4 Survey ahead of retrofit

The drive towards making traditionally constructed buildings more energy efficient will see an increasing number of surveys taking place ahead of proposed retrofit work of both internal and external insulation. This will require consideration of both the technical and aesthetic characteristics of brickwork as discussed in Chapters 3 and 4 and therefore all stages of the survey methodology

Fig. 1.8 Terraces of brick houses such as this are often seen as suitable for external wall insulation when large-scale retrofit is planned. Before retrofit work takes place, however, a thorough understanding of the technical and aesthetic characteristics of the structure should be undertaken, especially in relation to moisture movement and detailing of insulation at reveals.

set out below will be relevant. The use of external wall insulation on buildings of traditional brickwork will, in many cases, be inappropriate for both technical and aesthetic reasons. Technically, any insulation should be compatible with the existing brickwork. For buildings of traditional construction, this will often mean the preservation of moisture movement through bricks and the insulation material. Aesthetically, it is unlikely that the application of external wall insulation will be appropriate for traditionally constructed brick buildings. If a survey is being conducted ahead of retrofit, therefore, a thorough assessment of the heritage significance, character and material properties of the brickwork should be carried out. The condition of brickwork should also be assessed where retrofit is being proposed. Where any signs of decay or damage are present, this should be investigated further as brickwork should be in good condition if any insulation, whether internal or external, is proposed. It will be apparent from this brief consideration that insulating traditional brickwork requires a thorough assessment of heritage, technical, aesthetic and condition considerations, and should not be undertaken until this has been carried out. Retrofit of traditional buildings is a complex area, and further information can be found in Jenkins and Curtis (2021).

1.2 Executing the survey and assessment

1.2.1 Indicative survey methodology

The methodology and focus of any survey of traditional brickwork will depend on the aims and objectives of that survey, as discussed in section 1.1. As a broad generalisation, the survey of traditional brickwork can be broken down into three distinct stages.

- *Stage 1:* An initial survey to understand the brickwork and its context. This will include identification of how brick is used – for example, solid wall, arched flooring etc. Basic technical features and aspects of construction, such as bond and gauge, will be identified here. The use of decorative features will also be recorded at this time. This base of technical and aesthetic knowledge will inform further work related to condition.
- *Stage 2:* The second stage of survey examines the condition of brickwork. This will include identification of defects in bricks, mortar and the wider structure. Such defects are discussed in detail in Chapter 6 and include, at the most basic level, everything from open mortar joints and decayed bricks to larger structural failure.
- *Stage 3:* Having identified the technical and aesthetic characteristics of brickwork and the presence of any defects, it may be necessary to undertake specialist investigation or analysis. The various techniques that could form part of this work are discussed in detail in Chapter 7 and include mortar analysis and analysis of sulphates or surface coating.

There is also a need to then present the results of survey and assessment of traditional brickwork, which may be considered a fourth stage; this is considered in section 1.4. The focus of any survey will depend upon the purpose to which the results will be put, as set out in section 1.1. If, for example, the survey is designed to ascertain heritage significance, this will see a greater emphasis on stage 1 than stage 2. If the purpose is to identify potential repair needs, the emphasis is likely to be on stage 2 (although stage 1 will still be of import).

The methodology set out concisely above contains, as its central principle, the idea that when surveying a traditionally constructed brick building, it is vital to fully understand the brickwork that is being assessed prior to consideration of condition or any defects that may be present. This requires consideration of the technical characteristics of the brickwork as set out in Chapter 3. It will be seen in this chapter that a survey of the technical features of brickwork includes assessment of the bricks themselves, the bond and gauge to which the masonry has been constructed, the way the brickwork has been built and other factors such as the presence or otherwise of damp-proof courses, reinforcement and arches. A thorough understanding and recording of these technical features is the basis for subsequent work to identify defects and their probable cause. An understanding of the technical characteristics of brickwork is therefore the foundation upon which subsequent survey work will be built.

Following a thorough assessment and recording of technical aspects of the brickwork, it is necessary to consider decorative features that may be present. These are described more fully in Chapter 4 and include polychromatic brickwork, dogtooth and dentil courses, and the use of glazed bricks. These decorative features are also important to record as they add greatly to the character and appearance of many traditional brick buildings. When considered in conjunction with technical characteristics, they can also provide important indicators regarding the age of brickwork.

This naturally leads to an assessment of the age of brickwork and its heritage significance. The broad development of brickwork in the United Kingdom is discussed in Chapter 5 along with the technical and decorative characteristics that can help identify the age of brickwork. The basis when attempting to date a piece of brickwork, and ascertaining its heritage significance, is often knowledge of the use of brick in a particular country or region. Where assessing the age or heritage significance of a piece of brickwork is the primary aim, any survey work will need to first identify both technical and aesthetic features of that brickwork. This will make it possible to place the brickwork into its proper context within the overall technical development of traditionally constructed brickwork. Certain technical features will help with the dating of brickwork, as discussed more fully in Chapter 3 and in Case study 3.8.

When surveying the condition of a piece of brickwork, an examination of a structure with respect to the defects identified in Chapter 4 will be most appropriate. This will allow the building professional to both identify the symptoms of defects within bricks and brickwork and consider their likely cause. These may be relatively easily identified, such as spalled bricks as a result of long-term blockages to downpipes. Alternatively, they may be complex and the cause hidden from easy view, as may be the case with decaying bond timbers.

The surveying of traditionally constructed brickwork should include a broad assessment of a range of characteristics. In some cases, additional,

Fig. 1.9 The technical characteristics of this brick structure are complex and will also require consideration of the use of stone; a number of defects are also present.

Fig. 1.10 Traditionally constructed brick structures can present some intriguing and
challenging aspects to survey and assess, as can be seen in this 19th century
warehouse building.

specialist investigation or laboratory analysis may need to be carried out. This
may involve the use of specialist equipment on the part of a building professional or in some cases the use of specialist companies to undertake detailed
analysis or investigation. Techniques of this sort are considered more fully in
Chapter 7.

When surveying traditional brickwork, it is always important to note the
limitations of what can be seen, especially from a visual inspection. As noted by
Watt, it should be remembered that when a building is inspected visually, less
than 10 per cent of the fabric may be available for direct observation. Much of
a traditional building's fabric may be covered by external finishes, such as paint
or render, or internal finishes, such as lath and plaster, or may be unobservable
due to difficulties with access. Despite this, the building fabric that is visible
will often give indications of invaluable information regarding technical and
aesthetic characteristics, methods of construction, and decay and defects that
may be present.

1.2.2 Identifying the use of brick

One of the challenges when assessing brickwork is often as simple as identifying how brick has been used. Brick is found in many different applications in the historic built environment. It can be found as the principal masonry material in buildings of almost all ages and types. As set out in the introduction, a brick is defined as a unit of fired clay, which is generally of a size that can be held by a bricklayer in one hand. This type of modular fired clay masonry has been used throughout the world for much of human history. For buildings that employ brick as the principal masonry material, seeing this on the outer face of a façade is an easy enough task. However, there is no guarantee that the brickwork behind what is seen on the façade is the same type of brick, laid to the same standards using the same gauge or bond. Common brick could be used as a backing to facing brick, for example. Likewise, where brick is used as an internal partition, this will often be covered by lath and plaster, or later plaster board. Brick has also historically been employed in conjunction with a wide

Fig. 1.11 Brick is often found used to form the rear and side elevations of buildings with a front façade of stone, as seen here.

range of other materials. Brick arches, for example, may be used in buildings where stone is the principal masonry material. Examples of this can be seen in Figure 1.12. It is beyond the scope of this work to list the myriad ways in which brick has been employed around the world; what is essential in any survey or assessment is to clearly identify how and where brick has been used, to record this accurately and to gain access to as much of the brickwork in a structure as possible. The diversity of uses to which fired clay bricks have been put would fill several books alone. Good reference works on this subject include Brunskill (2009).

Fired clay bricks have been used in an exceptionally diverse range of ways in traditional buildings. This is seen most obviously in the construction of solid masonry walls – solid masonry referring to a wall constructed without a cavity. Walls constructed of brick can be exceptionally strong and durable. In common with other masonry types, brick walls perform well in compression and are therefore well suited to traditional construction practices.

The ways in which brick has been used in construction have never remained static over time. Technological advances and changes in fashion and design preferences have all influenced how the material has been used. As a result

Fig. 1.12 Brick can be used in a wide variety of ways in conjunction with stone walls; here the arches above the opening in this agricultural building are formed of brick.

of a more thorough understanding of how masonry behaves, coupled with advances in technology such as the more extensive use of cast and wrought iron, brick in the 18th and 19th centuries came to be used in increasingly innovative ways. This can be seen in the use of brick to form industrial buildings for various purposes. Brick was also extensively utilised in civil engineering works. This is most commonly seen around railways, where bridges and tunnels were often constructed of the material. Other structures, such as tall chimneys and lighthouses, were also built of brick.

When surveying any traditional building, it should be noted that brick may be found used in conjunction with other masonry types. This is discussed more fully in section 3.7.3. Brick could be used as a backing material for stone; therefore, a building which may present as a stone structure could, in fact, be largely built of brick. The tenement in Figure 1.11 seems to be constructed of stone; however, the side and rear elevations are constructed of brick, as are the internal walls. Some 80–90 per cent of this seemingly stone built structure is therefore fired clay brick. One of the earliest stages in surveying traditional brickwork is therefore to accurately record the ways in which brick is used and the purpose to which the material has been put.

1.2.3 Measured survey of brickwork

One of the fundamentals of surveying brickwork is often gathering dimensional data. This may take the form of measuring whole buildings, parts of a building or a particular area which requires intervention. The scope of this work precludes detailed consideration of this although a number of works in the References and Further Reading section at the end of this chapter, most notably Swallow et al. (2001), give detailed guidance on carrying out these functions.

When assessing traditionally constructed brickwork, the most common measurements that are required are generally to ascertain the size of bricks, the height and width of mortar joints and the gauge to which the brickwork was constructed. These are generally relatively easy measurements to make using conventional equipment, with the caveat that the scale of any measuring equipment should be sufficient to accurately record relatively small differences in brick sizes or joint height and width.

Recent developments in digital documentation and related techniques may be useful in some aspects of the survey of traditionally constructed brickwork, but as these are unlikely to be applied to the majority of traditional brickwork, they are discussed in section 7.10.

Measurement can also be critical in ascertaining the presence of defects. Measurement of brickwork can reveal, for example, whether parts of a structure are out of alignment with other parts around them. Careful and correct measurement during survey work is therefore critical in terms of both understanding the development of the structure and investigating any defects. Changes in the size of bricks or the width and height of mortar joints can be indicative of different phases of construction. They can also help to identify later interventions – for example, where a wall is constructed of Victorian imperial-sized bricks, later patch repairs using metric brick sizes will readily be noted during measurement work. Carefully measuring brickwork, therefore, is an integral part of any detailed survey and assessment of a traditionally constructed building.

Fig. 1.13 Measured survey of traditional brick structures can involve the assessment of complex structures such as this 19th century industrial building.

Fig. 1.14 At the other end of the spectrum from Fig. 1.13 is the measurement in survey of specific features such as the width of this pilaster, or indeed the measurement of a single brick.

1.2.4 Access

One of the most significant challenges when surveying traditionally constructed brickwork is often gaining access to ascertain technical and aesthetic features as well as assessing condition. While much can be seen regarding type of brick, the way in which it is used and condition from ground level, this leaves a lot of brickwork that has not been assessed. The aid of binoculars can assist with inspection of higher-level brickwork, but it must be acknowledged that brickwork that is hidden from view – for example, chimney stacks – will be difficult to properly assess from ground level. The use of scaffold towers, cherry pickers or other access equipment may be required in such a circumstance. Specialist access in the form of rope access inspection may also be required. The use of new technology, such as aerial photography, may also be appropriate in some instances, with the caveat that relevant regulations and legal requirements will impact the ability to use this technology. The level of access required, and the techniques employed to do so will be specific to the requirements of each building being assessed. Health and safety considerations should always be paramount, and there may also be other legislative or regulatory requirements around gaining access to inspect brickwork. Where brickwork cannot be accessed for whatever reason, this should be carefully recorded as it may be something that needs to be looked at in the future.

Fig. 1.15 Engineered structures frequently utilise brick in their construction, such as this bridge; they can present particular challenges with regard to access.

Fig. 1.16 Brickwork on domestic buildings can also present access challenges, especially at high levels such as this brick chimney stack.

1.4 Recording and presenting survey and assessment results

1.4.1 Reporting the results of survey and assessment

Regardless of the reason why traditional brickwork has been surveyed and assessed, there will be a need on completion to present the results of that work. Conventional reporting of the results of survey and assessment of traditional structures are likely to include a basic site plan, photographic record of the parts of the building examined and a description of what was found. The methodology discussed above provides an outline of what may be presented when reporting on traditional brickwork: the technical and aesthetic features of the brickwork; its age and significance; and defects and their cause. This is with the caveat that the balance between these and the level of detail will vary depending on the purpose of the assessment.

Almost as important as what was inspected is recording what was not. Any parts of a structure that were inaccessible, could not be assessed or were purposely excluded should be highlighted. Where defects are found in brickwork, these should be highlighted along with any evidence of their possible cause. In some cases, this may be relatively clear – as with a blocked downpipe – while for others the cause may be harder to discern. If there is no scope for further investigation, the need for this should also be highlighted.

The use of annotated photographs, as in Figure 1.17, can be a useful way of presenting the results of survey work. Any photographs presented should be accompanied by clear details regarding where in a building they were taken. When material is being removed for analysis (such as mortar analysis), the location from which it was taken should be recorded clearly on a site plan and presented in any report.

Fig 1.17 The use of annotated photographs can be a helpful method for displaying the results of survey and assessment. Here, the need to replace bricks is highlighted in red. The original structure can be seen in the image on the left.

The results of specialist investigations are likely to be summarised in the main body of a report, with the full results of this analysis presented as an appendix. As discussed in Chapter 7, it is important to set out the aim of such specialist investigation, the methodology used and the results obtained. This may be followed by a discussion of the implications of the results of such testing. Where reports are presenting the results of survey and assessment work, this should be presented in a way that allows the original purpose of the survey to be fulfilled, whether this is identifying heritage significance, repair work, retrofit or a combination of these.

1.4.2 Building information modelling (BIM)

Building information modelling (BIM) is a broad term used collectively to refer to a range of digital means to create, manage and utilise information related to a building, facility or structure. Although often associated with new build construction or large facility management operations, BIM is increasingly coming to be used in many aspects of managing traditional and historic buildings. Various terms are used to describe the specific application of BIM to traditional buildings; the term historic BIM (sometimes abbreviated to HBIM) is generally accepted for this.

The use of BIM within traditionally constructed buildings requires detailed understanding of both the application of BIM and of traditional construction. This is largely beyond the scope of this work, but excellent works are referred to in the References and Further Reading section at the end of this chapter, with Frost (2018) being a good introduction.

BIM can be used for various purposes in relation to the survey of traditional brickwork. BIM can be used for recording and documenting the results of a survey. Depending on the survey methods being employed, BIM can also be a tool for the long-term monitoring of buildings and can therefore be part of the survey methodology. BIM can also be particularly helpful in coordinating the outputs of several specialist elements of a survey.

BIM allows for a disparate variety of information to be gathered together and held centrally. This can include geometric data relating to a building, condition reports and the results of specialist testing and investigation. A BIM model can therefore become a central focus for disparate information. It will be seen throughout this work that a BIM model that records interventions could be invaluable when trying to ascertain if a previous intervention has induced decay. If case study 6 had a BIM model that recorded previous cleaning of the brickwork and the type of surface coating applied subsequently been available, much speculation as to the cause of the decay could have been avoided.

The BIM process is perhaps most succinctly defined by Antonopoulou (2017, p. 8) as follows: "the BIM process involves the assembly of 'intelligent' objects (building components and spaces) into a virtual representation of a building or facility … A BIM consists of: geometry (2D and 3D); non-geometric information; linked documents and data". When the model is thought of in this way, it quickly becomes apparent how the various aspects of surveying traditional brickwork detailed in this volume can coalesce in the context of BIM.

Gathering the data to enter into a BIM can involve the use of a wide range of survey techniques. As a digital means of presenting information, it will be immediately apparent that digital means of gathering information about a building, including laser scanning and photogrammetry, will mesh easily with BIM. Yet this does not mean that conventional methods of ascertaining data, such as the measuring of brick sizes or gauges, cannot fit with a BIM.

BIM clearly represents an opportunity to present the results of the various surveys, assessments and analyses of traditional brickwork described in this work in an easily accessible way. While not applicable to all structures, BIM is likely to play an increasing part of the survey of traditional brickwork in the future.

1.5 A conservation–based approach to survey and assessment

When carrying out any work to traditionally constructed brickwork, it is important to take what may be described as a conservation-based approach. This can be considered the application of conservation philosophy to the survey and assessment of traditional brickwork. "Conservation philosophy" is the term used to describe the guiding principles that inform work to traditional and heritage buildings. There are many versions of conservation philosophy set out by national and international heritage organisations, which are referenced at the end of this chapter and in the Bibliography and include examples such as the Burra Charter. The British Standard BS7913:2013 *Guide to the Conservation of Historic Buildings* is a recent example of relevance to those working in the United Kingdom. It sets out conservation principles in a relatively straightforward way, making it a good starting point when seeking guidance on conservation philosophy in surveying historic and traditional brickwork. Central to this standard is the assessment of significance, which is generally connected to the age of brickwork and the survival of original fabric or craft techniques.

A comprehensive examination of conservation philosophy and how this relates to surveying traditional brickwork is beyond the scope of this work. To give an indicative example from section 7.5 of BS7913 states that, "Materials selected should be of appropriate quality, suitable for the intended use and sourced for the particular historic building to achieve best performance match as well as best aesthetic match." A conservation approach is urged when dealing with historic buildings; the standard states that, "The removal of historic fabric and patina should be avoided as far as possible to retain authenticity." The principle of significance features prominently within the standard: "Those directing the works should understand the significance of the historic building." A checklist is provided to aid in the identification of significance; while not directly related to brickwork, this is a useful starting point. Meeting the requirements of these clauses presupposes an accurate identification of the existing materials and their technical and aesthetic features.

Fig. 1.18 A conservation-based approach to survey and assessment can help to retain as much original fabric of traditional brick structures as possible.

Chapters 3 and 4 of this book seek to provide a firm foundation of knowledge and understanding to contribute towards this where brickwork is being assessed. The depth of knowledge that a professional will require in relation to conservation philosophy will depend on the building being surveyed, but a basic understanding of its principles is a good foundation for any work to traditionally constructed brickwork. Placing conservation at the heart of any survey or assessment work will help to ensure subsequent interventions are proportionate and aid the retention of significance.

1.6 Case study 1: Susannah Place, Sydney, Australia – a conservation-based approach in action

Thanks to Nicola Ashurst, Adriel Consultancy for contributing the material for this case study.

This case study discusses a survey of a terrace of four houses and a corner shop in the inner-city area of Sydney known as The Rocks, built in 1844 and adjacent to Sydney Harbour. The brick and sandstone houses feature basement kitchens and backyard outhouses. The buildings, which today house a museum on working-class history, have survived numerous demolition threats – in 1900, when a Bubonic plague led to hundreds of neighbouring properties being razed; in the 1920s, when a three street-wide section of The Rocks was cleared to make way for the approaches to the Sydney Harbour Bridge; and in the 1970s, when labour disputes halted numerous demolition and redevelopment projects that would have seen many historic sites such as this lost forever. Susannah Place features the only surviving working-class dwellings of its era and is unique in having a history of domestic occupancy from its construction until 1990. Susannah Place is a protected heritage site and was added to the New South Wales State Heritage Register on 10 December 1999. Today it is known as Susannah Place Museum. The museum aims to preserve the evidence of the building's use and its adaptation to the changing needs of its occupants over 150 years.

The building is constructed on a sandstone foundation, with external brick walls in colonial bond and internal walls brick nogged. The roof line is hidden by a sandstone-capped parapet, in the fashion of the day. Each house was originally built with six rooms over three levels, with the kitchen in the basement and external outhouses. Susannah Place is very similar to terrace houses in the United Kingdom that featured a simple pattern of openings and detailing. It is outstanding for the quality of its external brickwork.

Sydney played an important part in the development of brickmaking in Australia. The penal settlement at Port Jackson (Sydney) was established on 26 January 1788. Within six months, it had been noticed that there was plenty of clay available for making bricks and a considerable quantity of them had been made and fired by that time: "Brickmaking was an important activity in the early colony and essential to its survival as the local timber and stone were not suited to the British building practices" Tench (1979, p. 5). Bricks at this time were fired at low temperatures and made with unskilled labour. There was a major centre of brick production at Brickfield Hill, near what is now Central Station.

The external brickwork of Susannah Place is comprised of low-fired, highly porous bricks bedded in wide beds of shell lime mortar. The external wall surfaces have a long history of tinted limewash application. There is a high level of compatibility between these materials; they work together outstandingly well. Intervention in the walls has been minimal, and as a result they were found in survey work to require virtually no conservation or repairs after 176 years.

When the building was surveyed, it was concluded that no work was required to the brickwork, even in areas of heavier weathering. There was also found to be a symbiotic relationship between the three super porous materials of which they are constructed. When considered in the context of the local Sydney climate, which is notorious for soluble salt crystallisation, damage due to the marine aerosol plume which extends 5 miles (8 kilometres) inland as the wind blows, this lack of decay or damage is all the more remarkable. Any evidence

Fig. 1.19 The Gloucester Road (front) elevation of Susannah Place (copyright Nicola Ashurst/ICS, Sydney)

Fig. 1.20 Some of the cracks to the East (rear) elevation, repaired in lime mortar (copyright Nicola Ashurst/ICS, Sydney)

of salts is barely noticeable. All the materials have coped with it admirably. The brickwork has some cracks due to structural movement, but these were not considered to require intervention.

This case study demonstrates the importance of a conservation-led approach to surveying and assessing traditional brickwork. It is clear from the assessment of the brickwork contained in this case study that, despite the age and many changes to which the brickwork has been exposed, no defects or intervention were required. The assessment of the brickwork did record a number of

Fig. 1.21 The beautiful quality of the untouched 176-year-old brickwork. Some joints still retain their original, slightly weather-struck pointing profile (copyright Nicola Ashurst/ICS, Sydney)

Fig. 1.22 Close-up view of the original mortar, a burnt shell lime mortar that used oyster shells. The mortar has much evidence, which suggests it was a hot mix – although that is not conclusive. The sand is an evenly graded beach or dune sand. (copyright Nicola Ashurst/ICS, Sydney)

technical and aesthetic features of significance, most notably the joint profile applied to mortar joints and the use of traditional limewash as a surface coating. This case study highlights the importance of correctly understanding brickwork that is being surveyed and of taking a conservation-led approach to assessing and recording that brickwork.

References and further reading

Antonopoulou S.B. (2017), BIM for Heritage: Developing a Historic Building Information Model, London: Historic England.

British Standards Institute (2013), *BS 7913:2013 Guide to the Conservation of Historic Buildings*, London: British Standards Institute.

Brunskill R.W. (2009), *Brick and Clay Building in Britain*, New Haven, CT: Yale University Press.

Curtis R. and Hunnisett J. (2017), *Climate Change Adaptation for Traditional Buildings*, Edinburgh: Historic Environment Scotland.

Earl J. (2003), *Building Conservation Philosophy*, Shaftesbury: Donhead.

Frost A. (2018), *Short Guide: Applied Digital Documentation in the Historic Environment*, Edinburgh: Historic Environment Scotland.

Glover P. (2006), *Building Surveys*, Oxford: Elsevier.

Jenkins M. (2014), *Scottish Traditional Brickwork*, Edinburgh: Historic Scotland.

Jenkins M. and Curtis R. (2021), *Retrofit of Traditional Buildings*, Edinburgh: Historic Scotland.

McCaig, I., ed. (2013), *Practical Building Conservation: Conservation Basics*, London: English Heritage.

Oxley R. (2003), *Survey and Repair of Traditional Buildings: A Sustainable Approach*, Shaftesbury: Donhead.

RICS (2018), *Surveying Safely*, London: RICS.

Sowden A. (1990), *The Maintenance of Brick and Stone Masonry Structures*, London: Spon.

Swallow P., Dallas R., Jackson S., and Watt D. (2001), *Measurement and Recording of Historic Buildings*, Shaftesbury: Donhead.

Tench W. 1789 [1979]. *Sydney's First Four Years* (facsimile edition), Sydney: Royal Australian Historical Society in association with Library of Australian History.

Warren J. (1999), *Conservation of Brick*, London: Butterworth-Heinemann.

Watt D. (2007), *Building Pathology: Principles and Practice*, Oxford: Blackwell.

Watt D. (2011), *Surveying Historic Buildings*, London: Routledge.

2   Traditional bricks and mortar

Traditional brickwork is formed primarily of two materials: fired clay bricks and mortar. This chapter seeks to provide a foundation of knowledge around these two materials. In the first part of the chapter, bricks will be considered. The process of manufacturing bricks will be examined, as will the raw materials from which bricks are made. The material properties of fired clay bricks will also be discussed here. Following this, mortar will be considered. This will examine the binders used – primarily lime, but also early Ordinary Portland Cement – as well as aggregate and methods of using mortar to build brick buildings. As with other areas examined in this book, what is presented here can only be regarded as a limited summary of the many different types of brick and mortar used, not only throughout the United Kingdom but across the world. What is presented here should be seen as indicative of the knowledge and understanding required for those undertaking survey and assessment of traditional brickwork.

2.1 Traditional bricks

2.1.1 Raw materials for brickmaking

The basic raw material used in the manufacture of bricks is clay. There is, however, no one material that can defined as "clay" but rather a spectrum of materials with certain specific properties. The essential characteristics of such a material in the context of brickmaking can be defined as follows: the substance is a natural material with plastic properties or in which plastic properties can easily be developed; it is comprised of particles of a very fine grain size; that these particles be largely or wholly hydrous aluminium silicates (Eyles and Anderson 1946, p. 30). Contemporary sources such as Allen (1893) support this mineralogical composition, stating that the elements present in good brick clay are silica and alumina, the silica being in chemical combination with the alumina as opposed to a mere mechanical mixture.

The term "clay" is generally taken to refer to materials that are naturally moist and possessed of a degree of plasticity and softness. However, while some hard shales exhibit none of the plasticity that would be expected of clay, they can still be used for brickmaking due to their mineralogical composition following

DOI: 10.1201/9781003094166-3

mechanical processing. Clays were often interspersed with sand; in such a situation, they were referred to as sandy clays, loams or sandy shales. If too great a quantity of sand or silica was present in a clay, it was unsuitable for brickmaking as the brick would become too brittle. Conversely, if the material being used for brickmaking had too great a quantity of alumina, the brick would shrink and distort during firing.

The final element necessary in a good brick clay was a flux to aid the fusing of the alumina and silica together during firing. This varies depending on local geology – for example, in parts of England lime forms the flux but in Scotland, due to the geology of the clay deposits, ferric (iron) oxide often performed this function. If no such fluxing agent was present, the clay was known by brickmakers as "foul clay", as the elements would not fuse together properly during firing. A good clay for brickmaking was therefore one that contained silica and alumina together with just enough flux to fuse them together.

Historically, a wide range of clays have been used for brickmaking in the United Kingdom. To name but a few of the more important of these, clay from the Oxford Clay Formation was used to manufacture Fletton bricks; carboniferous clay was used in the North of England and central Scotland, where shale from this era was used to make what were termed "colliery" bricks; and surface deposits of boulder or alluvial clay were used throughout the United Kingdom. To give details of the many sources of clay used historically in brickmaking in the United Kingdom alone would fill several volumes (some works are referred

Fig. 2.1 Extraction of clay from a shallow surface deposit; a wide variety of sources of clay have been used in the United Kingdom and beyond for brickmaking (copyright Courtesy of HES).

to at the end of this chapter). What is important to note in the context of survey and assessment is that the use of local sources of clay can give traditional brickwork – particularly that constructed before railways allowed bricks to be transported considerable distances – a highly regional character, adding to the significance of traditional brickwork. The use of different raw materials also leads to differences in performance and technical characteristics.

Before clay was used for making bricks, it was often put though a mill to increase plasticity and remove any large inclusions. In the 18th and early 19th centuries, this would have been achieved by the use of a pugmill. When harder raw materials such as shale came to be used and new brickmaking processes such as stiff plastic and semi-dry were employed, traditional pugmill were not sufficient to prepare the tougher raw material and other machines were developed. These included pan mills and large roller-based crushers. Pan mills consisted of circular pans in which the clay was placed, and two large grinding stones or rollers crushed the clay.

2.1.2 Moulding the brick

Having extracted and processed clay, the next stage in the brickmaking process is to mould the clay to shape. Prior to the mid-19th century this was carried out by hand. There have historically been a number of variations of the hand moulding of bricks. Sand moulding (sometimes referred to as stock moulding) involved a piece of wood called a stock board being fixed to the moulding table onto which the mould box used to form the brick was placed. From the late 18th century, the stock board could have a raised block attached, which formed the "frog" of the brick. The stock board and the mould box were sprinkled with sand prior to a lump of clay sufficient to slightly overfill the mould box being cast in, the excess clay being removed using a "strike" or "harp" and the brick turned out ready to be removed for drying. Skilled moulders would generally be assisted by an unskilled worker, to ensure the clay was ready in suitably sized lumps.

In slop moulding, the mould was placed directly onto the top of the table at which the work was taking place, this being wetted rather than sanded to facilitate the removal of the unfired (often termed "green") brick. In some cases, slop-moulded bricks were formed directly on the ground, where they were allowed to dry prior to firing. It was generally considered that sand-moulded bricks were of a more regular shape, with sharper arrises than slop moulded bricks.

By the 1830s, a number of attempts were made to mechanise brickmaking. These early brickmaking machines mostly worked on the principal of extrusion. This is the process whereby a column of clay is forced through a die and then cut to the appropriate size by wires (hence the term "wire cut bricks" often being applied to the product of extrusion). Machines to make bricks and tiles came into relatively common use in the United Kingdom from the mid-19th century onwards.

Machinery to press bricks in order to increase their density and uniformity of appearance were developed from the mid-19th century. Initially, these were

Fig. 2.2 For the majority of the period that may be defined as traditional construction, bricks were made by hand. The hand moulding of bricks continued through the 19th and 20th centuries and is still practised by some brickmakers today (copyright Furness Brick).

Fig. 2.3 The 19th century saw considerable developments in the manufacturing process of brick. Large machines such as this were employed to both press bricks to shape and to manufacture them by extrusion. This resulted in a great variety of brick types being produced.

Fig. 2.4 The ability to press facing bricks allowed for durable, mass-produced special bricks to be used from the latter half of the 19th century.

small and operated by hand, generally being used to improve bricks first moulded by hand or extrusion. Later, these machines developed in size and complexity and were powered by steam. The clay used in such presses was of similar consistency and plasticity to that used in extrusion and this technique was therefore known as the "plastic press" process. Later, machinery that allowed the use of what were called the stiff plastic and semi-dry processes were developed.

The stiff plastic brickmaking process allowed harder clays to be used in brickmaking and the use of colliery shale as a raw material. In the stiff plastic process, clay or shale was ground to a powder, mixed with water to bring about the required plasticity and then shaped into bricks by machine. A later innovation allowed powdered clay or shale to be formed into a brick by submitting it to great pressure without the admixture of water (although it would have been slightly dampened to give a little cohesion). This was known as the semi-dry method and is thought to have originated in the Accrington area of England in the 1860s.

2.1.3 Drying and firing of bricks

Having extracted, processed and moulded clay into the requisite shape, the next stage in the brickmaking process is drying and firing. Drying of bricks was traditionally carried out either in a drying shed, in the open air or in a building with louvred sides known as a hack house. As continuous kilns were developed, the drying process was achieved in the chamber adjacent to that being fired.

When sufficiently dry, bricks were fired, either in clamps or kilns. A clamp is a large construction of unfired bricks with fuel interspersed throughout. They are regarded as fairly crude mechanisms for firing bricks but were capable of firing great numbers at any one time, often between 150,000 and 300,000 in a single clamp. A clamp was formed by interspersing layers of unfired bricks with layers of fuel such as coal or coke. The bricks fired in clamps could vary considerably in quality and were carefully sorted when cool enough to handle.

If not fired in a clamp, bricks were fired in kilns – many variations of which were used in the firing of bricks in the period of traditional construction. These are usually categorised by how they operate, the simplest subdivision being between intermittent and continuous kilns.

Intermittent kilns are the simpler and older form of the two. Intermittent kilns can themselves be subdivided into three categories, depending on their mode of operation: up-draught, down-draught or horizontal draught. The simplest and earliest form of intermittent kiln was one in which heat from firing rose up through the bricks being fired and out the top known as up-draught. Down-draught kilns were an advance on up-draught. In these kilns, heat and smoke

Fig. 2.5 Continuous kilns such as this, along with moulding by machine, allowed considerable expansion in brickmaking in the late 19th and 20th centuries (copyright HES. Reproduced courtesy of J R Hume).

Fig. 2.6 Following the firing of bricks, they were sorted into those suitable for exterior use, and those that could be used externally. This process continues in traditional brickmaking today (copyright Furness Brick).

rose to the crown of the kiln; and was then drawn down through the bricks being fired, through a perforated floor and out through a chimney. Such kilns allowed very close control of firing and a greater uniformity of heat compared with up-draught kilns. It is beyond the scope of this book, but the development of kiln technology is detailed in a number of works (e.g. Hammond 1990). One of the most significant advances in kiln technology came in the mid-19th century with the development of the continuous kiln. A wide range of such kilns were developed within the period under consideration in this work, the most common being Hoffman, Belgian and transverse-arch kilns. All share the characteristic of being able to be in continuous operation, the kiln being formed in a number of distinct chambers, the firing passing from chamber to chamber with combustion gases being used to dry bricks in advance of firing in the next.

2.2 Material properties of traditional bricks

An understanding of the manufacturing process of bricks as set out above is important to fully understand the material properties of the material when it is used in a structure. Every stage in the manufacturing process influences the characteristics of the bricks. The raw material used clearly impacts both the technical and aesthetic characteristics of bricks. As discussed in section 3.2.3, the colour of bricks is influenced by the clay used in their manufacture. So too is the performance of the brick in key areas such as water absorption and resistance to frost. Almost as influential as the raw material itself are the extent and manner in which that raw material is processed, either by hand or in a pug or other form of mill. If large inclusions are not removed during processing, these may result in bricks that are more vulnerable to decay as seen in Figure 2.8. If, however, clay with the correct mineralogical content is used, and this is processed well, it will give a strong and durable brick.

The method of forming the brick will also influence its material properties and characteristics. As will be seen in section 3.2.1, brick type – which is heavily influenced by the method of manufacturing – influences properties such as water absorption. Bricks made by extrusion, for example, tend to exhibit a higher water absorption rate than those manufactured using mechanised pressing, as demonstrated in the testing referred to in section 7.5.

The way bricks are fired also contributes to their material properties. If they are fired correctly at the right temperature in relation to the raw materials used, again a strong, durable brick will be formed. As a general rule, bricks fired at a higher temperature are stronger and more durable than those fired at lower temperatures, although this is not absolute. If they are under- or over-fired, again the bricks may have vulnerabilities to decay. The sorting process after firing should have removed bricks that were not suitable for use in a building but this may not always have been carried out successfully or alterations to buildings may have led to bricks classified as suitable for use internally becoming external as seen in Figure 1.2.

Depending on the raw materials used to manufacture bricks and the firing process of the bricks, many brick types will undergo a process of vitrification on their surface. This is the process whereby clay begins to fuse to form a glassy bond partially filling the pores on the surface of the brick. This makes the surface of the brick more resistant to water penetration and therefore more durable. It is this vitrified surface that is often termed the "fire skin" of the brick. The extent to which this tough outer skin of the brick forms depends on the various factors discussed above. It will be seen later in Figure 6.7 that some bricks have only a very thin fire skin. The loss of the fire skin through the actions of decay mechanisms discussed in Chapter 6 can have a serious impact on the long-term durability of a brick, as seen in Figure 2.10.

A number of other material properties of fired bricks are influenced by the interlinked factors of raw materials, processing, method of forming and firing. The compressive strength of bricks can vary considerably, with Hendry et al. (2015) noting this as being anywhere from 7 N/mm^2 for relatively low fired bricks to over 100 N/mm^2 for dense engineering type bricks of the type shown in Figure 2.7. Porosity is also linked to a complex interaction of the factors discussed above in relation to manufacturing. It will be seen from the testing referred in section 7.5 that porosity is not necessarily a sure means of judging durability. Some bricks can suffer decay and loss of surface material and still retain a relatively durable surface on the façade of the wall. Others such as those shown in Figure 2.10, are unlikely to be able to continue to perform durably in the long term, having suffered this form of decay. Specialist analytical techniques can help in ascertaining the material properties of traditional bricks, as discussed in Chapter 7, but it should always be remembered that even within the same building, considerable variation may be present in the bricks used and decisions regarding condition may come down to a brick-by-brick assessment, especially where building defects have led to spalling of brick. A conservation-based approach as laid out in section 1.5 should always be taken.

Fig. 2.7 These dense, smooth engineering- type bricks have been manufactured by machine pressing. This provides a stronger, more durable brick.

Fig. 2.8 These bricks manufactured by extrusion have large inclusions in the clay. This is likely to be due to poor processing of the clay and has contributed to the spalling that is aff ecting the bricks.

Fig. 2.9 The bricks in this building, where they have been affected by the actions of frost, have lost their durable outer fire skin. It can be seen that the softer interior of the brick has been exposed, which is often more vulnerable to decay.

Fig. 2.10 These bricks have suffered such severe deterioration that some have completely disintegrated. The black lower fired centre of the bricks can be discerned where firing has not been wholly successful.

2.3 Mortar and traditional brickwork

To bed bricks in a wall, mortar is required. Given the significant percentage of a brick wall that is formed of mortar (this can be as great as 15–20 per cent), it has a significant impact on both the strength and durability of brickwork as well as its character and appearance. When surveying traditional brick buildings, therefore, consideration of mortar is of critical importance.

Both historically and in modern repair work, a wide range of different mortars can be found in traditionally constructed brickwork. The factors that influence both how a mortar performs and how it is defined include:

- the binder used
- the size, type and colour of aggregate
- the way the mortar is mixed and prepared
- the presence of any additives.

At the most basic level, mortar is differentiated by the binder used, often defined in simple binary terms as either "lime mortar" or "cement mortar". This is a crude differentiation, however, and within this dichotomy there are a wide range of other definitions that those surveying and assessing a traditional brick structure may have to consider. Having considered the mortar that is used in a wall, this should also be assessed in terms of historic joint finishes (section 3.4) and condition (section 6.5).

Visually identifying the binder used in a mortar is a difficult task. For this reason, where it is necessary to definitively ascertain the binder used – for example, if it is felt the mortar is contributing to the decay of bricks, or where survey is taking place ahead of repointing – mortar analysis, as discussed in section 7.1, will be required to fully identify the binder used.

2.3.1 Lime mortars

The majority of traditional brickwork would have been constructed using a mortar that employed lime as a binder. Within the broad category of "lime", however, fall a wide range of different materials defined by their properties and characteristics. Some confusion may arise due to differences between traditional and modern terminology for lime. Lime is generally subdivided into hydraulic and non-hydraulic.

Non-hydraulic limes, also referred to as "fat limes" or "air limes", are produced by burning limestone containing a high proportion of calcium carbonate in a kiln at temperatures around 850°C. This drives off carbon dioxide held within the lime to produce quicklime or calcium oxide. When water is added in a process known as "slaking", this will produce lime or calcium hydroxide. Once mixed with aggregate and used as a mortar, non-hydraulic lime sets, or carbonates, by reacting with carbon dioxide in the atmosphere and once again becomes calcium carbonate. This process is often referred to as the lime cycle; more detailed discussion of how this process works can be found in the works referenced at the end of this chapter (Historic England 2011). If not exposed to carbon dioxide, then non-hydraulic limes do not carbonate; this means such mortars are not used where brickwork will remain permanently saturated.

Hydraulic limes, historically referred to as "water limes", are produced from limestone containing naturally occurring clay silica or alumina impurities. The limestone is burnt in a kiln at temperatures in excess of 900°C, which as well as driving off the carbon dioxide, reacts with the impurities to produce quicklime. Lime manufactured from such limestone requires two reactions to set when used as the binder in a mortar: a carbonation reaction with carbon dioxide, as occurs with non-hydraulic lime; and a hydraulic reaction from the impurities that allow the lime to set in the presence of water.

The degree of hydraulic set and its strength are determined by the amount of impurities contained within the limestone that forms the raw material and the temperature and duration at which they are burned. The production and use of hydraulic limes increased during the 19th century and these materials were extensively used in brickwork for civil engineering structures such as railway tunnels, embankments and tunnels. The addition of a pozzolan to a mortar mix could be used to impart a hydraulic property to the binder. Pozzolans are materials that contain reactive silica and alumina, which are finely ground and added to lime; ground brick is one material historically used as a pozzolan.

The classification of hydraulic limes has changed over time, which can result in some confusion in terminology.

The classification of hydraulic limes was traditionally as follows:

- limes slightly hydraulic
- hydraulic limes
- limes eminently hydraulic.

This evolved to become:

- feebly hydraulic
- moderately hydraulic
- eminently hydraulic.

The definition of hydraulic lime mortars changed again in the 20th century. The current European Standard subdivides hydraulic limes to differentiate between those that are naturally hydraulic and those in which hydraulicity is brought about by the introduction of additives. The sub-categories are "natural hydraulic limes", "hydraulic limes" and "formulated limes". Today, the British and European Standard for building limes (BS:EN 459:1) classifies the strength characteristics in terms of an NHL designation and a number indicating the minimum compressive strength of a test cube of mortar at 28 days. Natural hydraulic lime is currently available in three categories (NHL 2, NHL 3.5 and NHL 5), according to their compressive strength. It should be noted that this does not directly equate to the three historical divisions in hydraulic lime discussed above.

There are a number of material properties common to a greater or lesser extent to most lime mortars that have made them well suited to traditional brickwork. The ability of lime mortars to allow moisture to dissipate through joints lets moisture diffuse from brickwork, helping to protect bricks from damage caused by saturation. Where mortar utilises a binder such as Ordinary Portland Cement (discussed below), that does not allow this diffusion of moisture through joints, it can have a highly damaging effect on bricks. Lime mortars are also often possessed of a degree of flexibility, allowing them to absorb minor structural movement associated with the expansion and contraction that brickwork undergoes due to changes in temperature and humidity. Both these material properties of lime mortar vary based on the type of lime in a mix and other factors, but most lime mortars will give a degree of moisture permeability and flexibility.

2.3.2 Ordinary Portland Cement mortar

A binary distinction is often made between mortar that uses lime as a binder and that in which the binder is Ordinary Portland Cement (OPC). When considering traditional brickwork, this binary distinction is often not helpful, as there are several historic "cement" mortars that may be original to traditional brickwork. The development of what may be regarded as cement mortars began in the late 18th century. The first commercially available cement binder is generally regarded as Parkers Roman Cement, patented in 1796. Developments continued in the early part of the 19th century, with proprietary products such as Keen's Cement (patented in 1838) and Parian Cement (patented in 1846) just two of many. Joseph Aspdin patented Ordinary Portland Cement in 1824. Portland Cement was formed by mixing relatively pure limestone with clay at a ratio of 100 parts limestone and 45 parts clay. This was fired at a higher temperature than traditional lime burning, the materials being first blended then fired at around 1200°C before being processed further (Bennett 2005).

The resulting material, termed Ordinary Portland Cement, had an immediate impact on brickwork as it was relatively easy to use and quick to set. It was, however, expensive, which initially limited the extent of its use. There were also concerns raised regarding the compatibility of the material with bricks, Pasley (1826, p. 204) states that "I have observed several thin walls … built with common mortar, but pointed with cement, that a great number of bricks have become defaced from this cause". This is an effect that would be recognised by conservation professionals today.

Ordinary Portland Cement mortars are generally stronger than those that use lime as their primary binder and are less able to allow for the dissipation of moisture. This can result in moisture becoming trapped within brickwork, and the attendant decay caused by this can be significant. The general principal regarding traditional brickwork is that mortar should be weaker than the brick it beds. Where modern cement mortar is used in a wall originally constructed of mortar that used lime as a binder, this creates inflexibility and can result in cracking and spalling of bricks, as discussed in section 6.2.

Care must be taken even when assessing buildings that were originally constructed using Portland Cement or other 19th century cement mortars. Although these are likely to have been stronger than lime mortars used traditionally, they would not have been as strong or inflexible as modern cements. Mortar analysis can help determine the binder used, but care must be taken in interpreting the results and modern cement should not automatically be considered the correct replacement binder for a traditional brick wall that was originally built of early cement. This is to stray into the realms of repair specification; for detailed advice on this subject, see the References and Further Reading section at the end of this chapter, in particular Hunnisett and Torney (2013).

2.3.3 Aggregate

In order to form mortar, the binder discussed above is mixed with an aggregate along with enough water to make the mix workable and cohesive. A wide range of different aggregates have been used historically in the formation of mortar to bed bricks, depending on what was available locally and what purpose the mortar was being put to. A good aggregate for a lime mortar was generally noted as being "sharp" (consisting of angular grains) and free of contaminants such as salt and organic matter. Mortar analysis can be helpful when identifying the size, type and quality of aggregate in a mortar mix from a building.

Aggregate is also often specified in contemporary sources as needing to be "well graded". Grading is the term for the size distribution of grains within aggregate. Where most grains are of a similar size, this is termed "poorly graded"; where a relatively wide distribution of grain size is found, this is termed "well graded". Well-graded sand aggregate typically has grain sizes between 4 mm and 0.125 mm, with most grains being around the midpoint. The grading and size of the largest grains within an aggregate are dependent on the type of brickwork being constructed. It is generally noted that the largest grain size should be no more than one-third of the width of a joint between bricks. For some forms of brickwork such as glazed bricks and gauged brickwork, mortar with a very fine aggregate will have been used. Regarding the water used for mixing mortar, the general stipulation was that this should be free of contamination and that potable (fit for human consumption) water should be used.

2.3.4 Mix ratio

The mix ratio for binder/aggregate in traditional mortars is often debated. The 1:3 ratio that is often defaulted to would not necessarily have been used originally – or, indeed, be the most appropriate for repairs. Mortar analysis can aid in ascertaining the original mix ratio of a mortar, but this will not necessarily give a specification for repair work. Several modern studies have shown this (e.g. Lynch 2007) and historic sources demonstrate a wide variety in mix ratios for traditional brickwork. Generally, historic mortars were lime rich, which can be seen visually and determined by analysis. Surviving historic specifications show that this ratio could typically range from 1:2 lime:aggregate ratio to as weak as 1:8 on some internal walling where economy of materials was exercised. Likewise, the strength of the lime used and the type of aggregate can vary widely, depending on the age of the building, location and purpose. Ascertaining mix ratios is an important part of any work to develop repair strategies; again the References and Further Reading section includes many excellent works to inform such work, especially Gibbons (2003) and Lynch (1994). It may also be necessary to analyse bricks themselves to ascertain material properties such as density, porosity and water absorption as part of survey and assessment prior to repairs using specialist techniques discussed in Chapter 7.

2.3.5 Preparation and use of mortar

Understanding the ways in which mortar has been used in the formation of brickwork can be an important part of survey and assessment. Identifying craft practices during visual survey can be difficult. It is sometimes possible to observe particles of unslaked lime where hot mortars have been used (discussed below), but even this is difficult. If information regarding the method used in preparing and using mortar is considered important in an assessment of traditional brickwork, it is likely that mortar analysis will be required. Mortar is clearly an integral part of all traditional brick structures and it should not be downplayed or overlooked during survey and assessment. Further detailed information on mortar and brickwork is referred to at the end of this chapter. Within this book, assessing the technical characteristics of joints is considered in section 3.4, the condition of mortar joints in section 6.5 and mortar analysis in section 7.1.

Having mixed the mortar, there are several methods for using it when constructing brickwork. By far the most common is simply to use the mortar as a bedding material. A layer of mortar is spread on the upper face of the bricks in the course onto which a brick is to be laid, with mortar also being applied to the side and rear of that brick where it will abut others in the same course.

A further method of using mortar was called "larrying; it involved spreading a thick layer of mortar over the last course laid and sliding the brick above it into position, thus squeezing some of the mortar along so it would fill up the joint between that brick and the next. This was most commonly used in large engineered structures with thick walls.

Some 19th century sources refer to the use of hot mortar. Partington (1825, p. 363) defines the practice as follows: "take of unslaked lime and of fine sand, in proportion of one part of the lime to three parts of the sand, as much as a laborer can well manage at once, and then adding water gradually, mix the whole well together with a trowel till it be reduced to consistence of mortar. Apply it immediately, while it is hot." The use of hot mortar, whereby the quicklime was slaked together with the aggregate and the resulting mix used immediately to bed bricks, is one likely to be found in traditional brickwork.

The terminology can get confusing around "hot mortars", especially given recent revived interest in the craft practice. Many historic mortars were made from a mix of quicklime, sand and water. This could be achieved by two distinct processes. In the first method, a measure of quicklime was slaked within the sand to a crude dry hydrate and then turned over and mixed fully before being punched through a screen to remove inclusions. This would then be mixed with water to the desired consistency and allowed to mature, or "bank", over a few days before being reworked, or "knocked up", ready for use. This is defined by Lynch (2007b) as "sand slaking" and produces a hot-mixed lime mortar.

Alternatively, the quicklime and sand could be mixed together as the lime slaked to be used as what is termed a "hot lime mix" mortar. As the process of slaking lime produces heat, the process is known as a hot mix. Hot mixes tended to be used soon after mixing for building work. However, as lime particles continue to slake, they expand, which can be problematic for relatively thin brick walls. This differs from a hot lime mix, which in the context of traditional brickwork sees water added over the quicklime and mixed straight away into a full mortar for immediate use. Its use in the context of brickwork was generally restricted to foundations and footings and other thick walls.

Lime putty (also called fat lime or slaked lime) is produced by slaking quicklime in an excess of water. Lime putty is fully slaked and typically allowed to

Fig. 2.11 A well-graded aggregate, consisting of a range of grain sizes ready for use in a lime mortar mix (copyright HES).

Fig. 2.12 Quicklime being slaked with water to produce lime putty (copyright HES).

Fig. 2.13 Lime putty ready for use (copyright HES).

Fig. 2.14 Slaking lime in damp sand (copyright HES).

"fatten up" for at least 48 hours prior to use in a lime mortar. The maturation or "fattening up" of putty results in the formation of increasingly finer lime particles over time. Some applications of brickwork used specific mortar types and mixes. Glazed brick, for example, used a mortar mix sometimes referred to as bricklayers' putty. This was a mixture of fine white sand or marble dust and pure lime, which has been slaked in a large quantity of water, strained and allowed to stand, commonly used in setting glazed bricks. Gauged brickwork also used a specific form of mortar; further detailed information regarding this can be found in Lynch (2007a).

2.4 Case study 2: Survey for heritage – recording wigging in Dublin, Republic of Ireland

With thanks to Graham Hickey at Dublin Civic Trust for sending me the original report on which this case study is based and the authors of the original report, Shaffrey Associates Architects, Dr Gerard Lynch and John Montague.

This case study presents the methodology used in a survey to assess the survival of a historic pointing technique known as "wigging" in Dublin, Republic of Ireland. Wigging is a distinctive Irish pointing technique that emulates the visual aesthetic of English tuck pointing. It differs from English tuck in that the ribbon and stopping mortar are a homogenous material formed in a single application and a colour mortar is then applied to give definition to the ribbon. A unifying "colour washing" of façade brickwork was also commonly applied.

The survival of this practice has been threatened by a variety of factors: a tendency to adopt craft practices from England during repair work and a gradual erosion of skills and knowledge of traditional materials have meant that the use of this craft practice has declined significantly. This is a problem common to regional and national craft practices in traditional brickwork throughout the world. There is also a concern about the ability of professionals undertaking the survey and assessment of brickwork to be able to identify this regional practice. All these factors mean that the craft practice of wigging was at significant risk of being lost entirely. The aim of the survey work noted in this case study was to provide an evidence-based approach to the repair of traditionally constructed brick buildings in Dublin.

The research project had several distinct aspects to its methodology. Data was collected through fieldwork observation of buildings by experts in traditionally constructed brickwork, coupled with research of documentary and archive sources. The fieldwork was restricted to building façades that could reasonably be recorded and that were likely to yield evidence of wigging. The fieldwork observations undertaken in this research were carried out on a cross-section of buildings primarily located in central Dublin. The predominant finding from the fieldwork observation was that wigging was indeed common in the city. This is important in addressing issues around skills, materials and appropriate specification for repair work of brickwork going forward. While fieldwork over a wider area may have resulted in a greater number of examples being found, by concentrating survey work on buildings that met these criteria, resources could be used more effectively. Following this fieldwork, qualitative analysis was undertaken. This aimed at ascertaining the extent to which the craft practice had survived, and to prove that this craft practice was once widespread. Following this, a further stage of the research project utilised trial panels of wigging techniques to assess and understand the techniques observed in fieldwork.

It is beyond the scope of this short case study to go into detail about the findings of the project. Important information around the mortar used, the use of colour wash and the historic authenticity of wigging were all found during the fieldwork stage. This case study wishes to highlight the way structured survey and recording can be used to ascertain traditional craft practices in relation to brickwork. These findings can then be used to inform repair and maintenance work to ensure it is authentic to regional and national craft practices. Work of this type is of integral importance to the survival of craft practices around the world, ensuring that traditionally constructed brickwork retains its character, its aesthetic and its technical performance long into the future.

In the case of the survey work discussed in this case study, the aim was not to assess the condition of the brickwork or the presence of defects (although in some cases this was a by-product of the work). Rather, the survey work discussed here was designed to assess the survival of a particular craft practice. This fits the broad definition of survey and assessment as set out at the outset of this work. As with all survey work, a correct understanding of the aims and objectives was vital to its successful completion.

Fig. 2.15 This image shows the technique used to recreate the joint profile known as wigging (copyright Dublin Civic Trust).

Fig. 2.16 Here, the coloured mortar used to form the joint profile can be seen. The extensive heritage-led survey detailed in this case study helped inform the re-creation of a vanished craft practice (copyright Nolans Group).

Fig. 2.17 Here a remaining section of joint profile can be seen to survive. This was found during the heritage survey of brickwork noted in this case study (copyright Dublin Civic Trust).

References and further reading

Allen G., Allen J., Elto N., Farey, M., Holmes S., Livesey P. and Radonjic M. (2003), *Hydraulic Lime Mortar for Stone, Brick and Block Masonry*, Shaftesbury: Donhead.

Bennett B. (2005), "The Development of Portland Cement" in *The Building Conservation Directory*, Tisbury: Cathedral Communications.

British Standards Institute (2001), *BS EN 459–1:2001, Building Lime: Definitions, Specifications and Conformity Criteria*, London: British Standards Institute.

British Standards Institute (2002), *BS EN 13139:2002, Aggregates for Mortar*, London: British Standards Institute.

Bonnell D. and Butterworth B. (1950), *Clay Building Bricks of the United Kingdom*, London: HMSO.

Brocklebank I. (2012), *Building Limes in Conservation*, Shaftesbury: Donhead.

Brown A. (2017), "Hot Mixed Mortars", in *The Building Conservation Directory,* Tisbury: Cathedral Communications.

Chapman S. and Fidler J. (2000), *The English Heritage Directory of Building Sands and Aggregates*, Shaftesbury: Donhead.

Dobson E. (1850), *A Rudimentary Treatise on the manufacture of Bricks and Tiles*, London: John Weale.

Davis C.T. (1889), *A Practical Treatise on the Manufacture of Bricks, Tiles, Terra-Cotta, Etc.*, Philadelphia, PA: Henry Carey Baird.

Eyles V.A. and Anderson J.G.C. (1946), *Brick Clays of North-east Scotland, Part 1: Description of Occurrences.* Wartime pamphlet, Geological Survey of Great Britain. London: Geological Survey and Museum.

Frew C. (2007), *Pointing with Lime, The Building Conservation Directory*, Tisbury: Cathedral Communications.

Gibbons P. (2003), *Preparation and Use of Lime Mortars*, Edinburgh: Historic Scotland.

Hammond M. (1990), *Bricks and Brickmaking*, Aylesbury: Shire.

Hendry A., McCaig I., Willett C., Godfraind S. and Stewart J. (2015), *Historic England Practical Building Conservation: Earth, Brick and Terracotta*, Farnham: Ashgate

Historic England (2011), *Practical Building Conservation: Mortars, Renders and Plasters*, Farnham: Ashgate.

Hunnisett J. and Torney C. (2013), *Lime Mortars in Traditional Buildings*, Edinburgh: Historic Environment Scotland.

ICS (1898) *Brickwork, Terracotta and Tiling*, London: Wyman.

Larsen E.S. and Neilson C.B. (1990), "Decay of Bricks Due to Salt", *Materials and Structures* vol. 23, no. 1, pp. 16–25.

Lynch G. (1994), *Brickwork: History, Technology and Practice* London: Donhead.

Lynch G. (2007a), *The History of Gauged Brickwork,* London: Elsevier.

Lynch G. (2007b), "The Myth in the Mix The 1:3 ratio of lime to sand" in *The Building Conservation Directory*, Tisbury: Cathedral Communications.

Partington C.F. (1825) *The Builder's Complete Guide*, London: Sherwood Stephen.

Pasley C.W. (1826), *Practical Architecture*, Chatham: Royal Engineering Establishment.

Rivington (1905), *Rivington's Notes on building Construction volume 1*, London: Longmans.

Searle A. (1913), *Cement, Concrete and Bricks*, London: Constable.

Torney C., Schmidt A. and Graham C. (2020), *A Data Driven Approach to Understanding Historic Mortars in Scotland*, Edinburgh: Historic Environment Scotland.

3 Technical characteristics of traditional brickwork

When brickwork is being surveyed, it is imperative to fully understand the masonry that is being examined. There are a number of technical features of brickwork that should be noted during survey work. This includes bond, gauge and the use of arches. Only when the technical characteristics of brickwork are fully understood can assessment of the building's condition be effective. Furthermore, the specification of repair work that may follow assessment and survey is largely reliant upon this information. By understanding the technical characteristics of traditional brickwork, the kinds of inappropriate repairs discussed in section 6.10 are more likely to be avoided. The technical characteristics of brickwork can also contribute to the significance in a heritage context and can help in ascertaining a date when brickwork was constructed.

3.1 Bond

The term "bond" refers to the way in which bricks are laid and interlocked in relation to each other within a wall. When describing bond, the terms header (a brick with its head or narrow end to the front) and stretcher (a brick with its longer face to the front) are used. Bond is defined in technical reference works from the 18th century onwards in broadly similar terms and is perhaps most succinctly said to be "the arrangement of the bricks so that the joints in one course are covered by bricks in the adjacent courses, and continuous vertical joints are avoided" (Adams 1906, p. 59). It is defined by Allen (1893, p. 12) as "the very essence of sound work … the bricks must break joint on the solid surface of a brick beneath … A straight vertical joint between two bricks exactly over a similar joint of the course below is exceedingly bad construction, and an unsound, weak piece of work." The main purposes of a bond are to provide strength to a wall and influence appearance; it can also be related to economy of construction. A wall's strength and stability are provided by ensuring that there are adequate bricks tying into each course with sufficient lap to avoid, as far as possible, straight vertical joints throughout the build.

In order to bring about the breaking of vertical joints, "closers" are employed. These usually consist of a brick cut in half along its length (technically termed a

DOI: 10.1201/9781003094166-4

"Queen Closer") and are employed in alternate courses to ensure that the joints between brickwork in one course do not fall immediately above or below those adjacent. An alternative method of achieving the same goal is to use a three-quarter brick bat at the start of each course.

A wide range of bonds may be found utilised in traditional brickwork throughout the world. The most common are Flemish and English bond. These are described more fully in Table 3.1. There are often regional variations in

Table 3.1 Some of the principal bond patterns found in traditional brickwork

Bond type	Description	Photograph
English Bond	Alternate courses of headers and stretchers often employed for its strength.	
Scottish Bond (English garden wall)	One course of headers to every three, four or five courses of stretchers, primarily employed due to its economy.	
Flemish Bond	Alternate headers and stretchers in each course, often described as being the most aesthetically pleasing bond.	
Stretcher Bond	All bricks in each course are stretchers except where a header is used in alternate courses to maintain bond. Generally indicative of cavity walling.	

(*continued*)

Table 3.1 Cont.

Bond type	Description	Photograph
Header Bond	In this bond all the bricks are laid as headers, commonly used for curved walls.	
Herringbone Bond	This is one of a number of decorative bonds sometimes encountered. Bricks are laid at an angle of 45 degrees in opposite directions. This forms a strong interlocking effect.	
Irregular Bond	An arrangement of stretchers and headers in each course that does not follow a pattern but does break joint to a degree	

bond patterns, a further instance of where knowledge of local construction practices can be beneficial when surveying brickwork. For example, in Scotland, the unsurprisingly named "Scottish Bond" is the dominant craft practice, this being formed of three, four or five courses of stretchers between each course of headers. In other parts of the world, this would be referred to as a variation of English Garden Wall Bond. A further variation in bond terminology can be seen in the case study at the end of Chapter 1, where Colonial Bond is referred to in an Australian context.

When surveying traditional brickwork, the bond pattern employed should be recorded accurately. This will involve, firstly, ascertaining the disposition of headers and stretchers within the brickwork being surveyed. This will allow for the bond pattern to be identified. To take one example, the wall in Figure 3.1

Fig. 3.1 Half brick closers are used in this 18th century Flemish bond brickwork; the joints of this brickwork are ruled.

Fig. 3.2 A wide range of bonds may be found in survey work, with the most common shown in Table 3.1. In this example, what is termed Monk Bond is formed of two stretchers between headers in each course.

Fig. 3.3 In this building, a three-quarter bat is used to maintain bond rather than a half brick closer, this is a craft practice more common in some parts of the United Kingdom than others; where used, it should be noted in any assessment of bond.

presents alternating headers and stretchers in each course, meaning it is built in Flemish bond. The wall in the last row of Figure 3.1 is a vernacular building which, while still employing stretchers and headers to form bond, does not follow a recognisable pattern and may therefore be termed "irregular bond". In addition to recording the bond pattern the method used to break joint should also be recorded. The example in Figure 3.3 would be recorded as "a three-quarter brick bat is used to break bond in alternating courses". There may also be instances in which a wall is built in a particular bond but that there are areas of the brickwork that deviate from the predominant pattern. This is most commonly found where polychromatic decorative schemes are used as discussed in section 4.3. Straight vertical joints are sometimes found in these cases. The other common situation where areas of brickwork may vary from the overall bond pattern is where repairs have been executed with little cognisance for the original work. In all cases, areas that deviate from the overall bond pattern are worthy of note in a survey.

3.2 Brick type, size and shape

3.2.1 Brick type

When defining the construction of a brick wall, ascertaining the type of bricks used is of central importance. However, formulating a typology of brick in traditional construction is something that, in practice, is a complex process. Brick type is influenced by the method of manufacture, raw material used and the purpose to which the brick is to be put. There is sometimes a mistaken belief that a brick is a generic building material and that all are interchangeable when, in common with other traditional building materials, this is not the case. Just as stone differs considerably in appearance and material properties, there are also a number of different types of brick that can be found in common use in traditional brickwork. Broadly, these fall into six main types:

- handmade bricks
- common bricks
- facing bricks
- glazed bricks
- special bricks
- engineering bricks.

It will be seen that these definitions are derived from a combination of the use to which the bricks are put and their method of manufacture. Recourse may be made to Table 3.2 for visual reference to these principal types of brick. Note that this is not an exhaustive list, but provides a starting point for defining brick types during survey work. The complexity and diversity of the range of brick types found in traditional construction does not easily fit this simple division, however.

Firstly, there are national and regional variations in brick type. For example, London Stock Bricks are a specific type of handmade brick used in London and the South of England. Generally yellow or brown in colour, they were to an extent superseded by machine-made Fletton bricks from the end of the 19th century. Thus, when surveying brickwork in London and its environs, regional variations in brick type may lead to bricks being noted as London Stock or Fletton in addition to the more ubiquitous brick types contained in the table below. These are brick types that would be absent from other parts of the United Kingdom, such as Scotland, or other parts of the world.

There are also historical terms for bricks that have become obsolete. Malms, Marl Stocks and Gaults are all terms for brick types that would have been readily understood in the 19th century but today are not in common parlance in either surveying or construction apart from amongst specialists in traditional construction. Using these terms in reports arising from survey work can be helpful to add greater understanding, but it is likely that these historic terms will be supplementary to descriptions of type that are more readily understood by modern audiences. Despite these differences between modern and historic

Fig. 3.4 A combination of pressed facing and glazed bricks is used in the façade of this building to create an impressive visual effect.

Fig. 3.5 Two different brick types can clearly be seen here: a red facing brick (some of which are specials) at the door reveal and stock brick on the main part of the wall.

Table 3.2 A simplified typology of bricks

Type of brick	Identifiable features	Illustration
Handmade	Applicable to a wide range of bricks, and all those produced prior to the advent of mechanisation in the 19th century. Can be of irregular shape and variable in colour and texture (although some are very precise and regular), generally thinner than those made through mechanised processes.	
Common (also referred to as colliery or composition brick)	Common brick is the name given to any brick that does not lay claim to any special properties and is intended to be used where aesthetics are of secondary importance. These bricks vary widely in their characteristics. Some can be almost as dense and durable as engineering bricks, while others such as those known as composition or colliery bricks, formed by pressing clay mixed with shale waste from coal mines, can be weaker and more porous.	
Facing	Where suitable quality clay was available, along with appropriate moulding techniques and kiln technology, a brickworks could manufacture facing bricks for use on principal parts of a building on public display. As with stone-fronted buildings and on all forms of walling of aesthetic importance, the higher quality bricks were used on areas that would be visible. Facing bricks will be of a higher quality and more durable than most types of common bricks.	

(continued)

Table 3.2 Cont.

Type of brick	Identifiable features	Illustration
Glazed	Those bricks manufactured with a glaze applied. They were commonly used in circumstances where a high degree of cleanliness and hygiene or light refracting was required, or for aesthetic purposes. Glazed bricks came in a wide variety of types, colours and shapes. Precise in shape, enabling setting with very fine mortar joints, they could be employed to striking visual effect.	
Engineering	Engineering bricks were formed from high-quality clay and manufactured in such a way as to make them very dense, impermeable and able to withstand greater compressive pressures than other forms of brick. Available as best quality, Class A blue engineering bricks or red coloured Class B engineering bricks, they are highly durable and can be found on civil engineered structures or sometimes on the lower levels of brick buildings where the brickwork is more vulnerable to damage and decay.	
Special shaped brick	The use of special bricks straddles the divide between aesthetic and technical characteristics of brickwork as they can be employed for both purposes. Special bricks are discussed fully in sections 3.2.3 and 4.4.	

Table 3.2 Cont.

Type of brick	Identifiable features	Illustration
Rubbers	Bricks made from clay free from inclusion to provide a brick of even texture that could be cut and rubbed for use in gauged brickwork. Often manufactured oversized to allow for the reduction in size during the cutting and rubbing process and fired at lower temperatures than would be used for other bricks (image copyright Georgian Brickwork).	

terminology and regional and national differences, some generalisations can be made with regard to brick type.

Handmade bricks would seem to present a relatively easy type to define, simply by their method of manufacture. However, there were historically considerable differentiations in handmade brick. This should be no surprise given that all bricks were handmade until mechanisation began in the mid-19th century. For example, Salmon (1748, p. 1) lists four types of brick: red stock, grey stock, place bricks and cutting brick. Surveyor (1787, p. 14) lists ten types of brick in total, all of which were priced at a different rate: best marl stocks, second best, common marl, picked grey stocks, grey stocks, place bricks, paving bricks, red stocks, Windsor bricks, rubbers red or grey. As discussed above, it is unlikely that such historic brick types would be used in modern survey, and the definition "handmade brick" would be the most likely application of type. The above references do, however, illustrate the complexity in brick type just within those that are handmade.

To take some further differentiations in brick type, an engineering brick is one manufactured from a clay that, when burnt at high temperature, vitrifies to produce a strong brick with low water absorbency. Engineering brick is generally dark red or bluish. An early form of engineering brick in the 18th century was referred to as a "water brick", due to its resistance to water absorption rate. A gault brick is generally pale yellow or off-white in colour, made from clay from the Gault Formation. These bricks are relatively soft and close textured, and found in the South and East of England. A Fletton brick was manufactured in Oxfordshire by the semi-dry method, originally around the Fletton area near Peterborough.

Each brick type will present specific features to aid in its identification. Pressed facing bricks are identifiable by homogeneity of colour, smooth texture on the face and sharp, well-defined arrises (edges). Common bricks manufactured from colliery shale can be identified by their rough texture and variegated colour. These factors are presented more fully in Table 3.2. As highlighted above, the range of brick types is considerable, and knowledge of regional and historic practices will always be highly beneficial.

The importance of accurately defining the brick type in a structure fulfils two distinct purposes. Aesthetically, especially where bricks are being surveyed ahead of repair work, matching brick type is of considerable importance. It is in technical performance that brick type is of critical importance. Most bricks have a tough, outer fire skin following firing, which adds considerably to the durability of a brick. Some brick types do not have a fire skin, however – for example, soft rubbing bricks used in gauged brickwork are of a fairly uniform consistency throughout. Correctly identifying a brick is of this type will prevent interventions that may be detrimental to the brick – for example, abrasive cleaning. Other brick types are also highly vulnerable to a loss of fire skin. Common bricks manufactured from colliery shale, for example, tend to have a softer, less vitrified interior, meaning that once the fire skin has been lost, the bricks are likely to deteriorate quickly. Denser, more durable bricks such as pressed facing or engineering bricks are more likely to be able to withstand some loss of surface material without exposing a softer interior. A correct assessment of the type of brick is therefore vital in assessing whether a brick is likely to need to be replaced or if it can safely be left in situ, and also to inform future interventions.

3.2.2 Size of bricks

The size of bricks can vary considerably, depending on type and date of manufacture. Generally, brick sizes have conformed to several key principles. Bricks were manufactured to be able to be held in one hand by a bricklayer, with the other hand being used to spread mortar. In order to maintain accurate bonding, the length of a brick must be around twice its width plus the thickness of a mortar joint. This allows for easier bonding of brickwork. For these reasons, whilst exhibiting considerable diversity in size, bricks generally conform to a relatively restricted set of dimensions.

At several points in history, the dimensions of bricks have been regulated. Between 1571 and 1776, what were termed "statute" bricks were specified for manufacture in England. This laid down a standard size for bricks. The Brick Tax, introduced in the United Kingdom in 1784, in some cases led to the manufacture of larger sized bricks as the tax was levied per brick. Changes in manufacturing methods also led to changes in the size of bricks, with mechanisation bringing about, in general terms, a move towards larger bricks. As with gauge, considered below, the size of bricks can be seen to exhibit regional variations. This can be seen within the United Kingdom in the 19th century, where brick manufacturers in the Midlands and the North of England generally produced larger sized bricks, with those produced in Scotland being larger still. This contributed to the variations in gauge noted in section 3.3. In the 20th

Table 3.3 Indicative development of the size of bricks in UK brickwork

Time period	Typical sizes	Notes
18th century to early 19th century	8–9 x 3 x 2.5 (inches)	These bricks were handmade and can vary in size but are generally thinner than later bricks.
Mid-19th to mid-20th century	9 x 4.5 x 2.75 (inches) Northern Gauge bricks in this period 3.75 inches thick	As the brickmaking process became mechanised, larger bricks were produced.
20th century imperial	8 5/8 x 4 1/8 x 2 5/8 (inches)	The British Standard for Imperial sized bricks.
Modern metric sizing	215 x 102.5 x 65 (mm)	The new modern brick size is generally inappropriate for use in repair of traditional brickwork.

Fig. 3.6 A front façade of pressed facing bricks can be seen here block bonding with a side elevation of common bricks manufactured from colliery shale. The difference in size and gauge between the two walls can also be clearly discerned.

century, further attempts at regulating the size of bricks took place. A British Standard for brick sizes was introduced in 1936, followed by the introduction in the 1960s of metric-sized bricks. In general terms, bricks used in the latter part of the 20th century showed a trend towards being smaller in size once again, particularly after the introduction of metric brick sizes.

In terms of survey work, the size of brick used is a relatively straightforward measurement, but care should always be taken that it is measured accurately. Generally, the average size of 12 bricks over several courses is taken to give an accurate indication of the size of the bricks throughout. It is important to take measurements of all relevant dimensions – length, breadth and height – which will require the measurement of both stretchers and headers within a wall. It should also be remembered that it was a common practice to use a higher quality brick on the front façade of a building than on the rear and side elevations, so two or more brick sizes may be found in the same building. Lastly, in some forms of construction, most notably cavity wall and some forms of brick infill to timber frame, brick was laid on edge rather than on bed.

3.2.3 Shape, colour and surface texture

Shape, colour and surface texture are three further characteristics of bricks that need to be assessed and recorded during a survey of traditional brickwork. These features of bricks are important for the aesthetic of the building and also relate to the performance of the structure.

Shape

In their standard form, bricks are rectangular prisms. However, there are always situations where bricks of a shape other than this are required for both technical and aesthetic reasons. From a technical perspective, in all brickwork there is a requirement to form half-brick closers or three-quarter bats to maintain bond. Where arches are being built, there is a need for bricks cut at an angle to allow the formation of skewbacks in arches other than semi-circular and, in all but rough arches, a need for bricks to be tapered to form voussoirs. Bricks may also be required out with the standard rectangular prism for openings, returns, curved or angled brickwork and countless other purposes. The decorative use of bricks shaped out with rectangular prisms are discussed more fully in Chapter 4 and include the formation of dogtooth courses and cornices, and allowance for polychromatic decoration.

The two principal methods of forming bricks of a non-standard shape are to cut a brick to the form required or to mould it to shape prior to firing. The tool that would have been used to cut bricks throughout the period of traditional brick construction is the brick axe. This tool has been researched and re-created very successfully by Lynch (2007, pp. 45, 207). It resembles a double-ended bolster (the modern tool for cutting brick) with a blade width of around 5 inches (13 cm). There are known to have been two styles of the tool, a smaller version that would have been in use in the 17th and 18th centuries and a larger, heavier version in use by the 1820s. When surveying traditional brickwork, it may be possible to see cutting marks where a brick axe has been used to cut bricks.

A brick that is purposefully moulded before firing into a shape that deviates from the standard rectangular prism is commonly referred to as a "special". Raw clay can be moulded into almost any shape to suit a particular purpose, and traditionally a wide range of bricks of different shapes were manufactured in order to fulfil standard architectural features, or bespoke bricks made for specific functional and/or aesthetic purposes. For example, the Scottish Terracotta Company manufactured over 300 different shapes of bricks in 1897. Such special bricks were sometimes manufactured of clay that produced a range of coloured bricks to allow their use in polychromatic decoration. Some of the most common special bricks are:

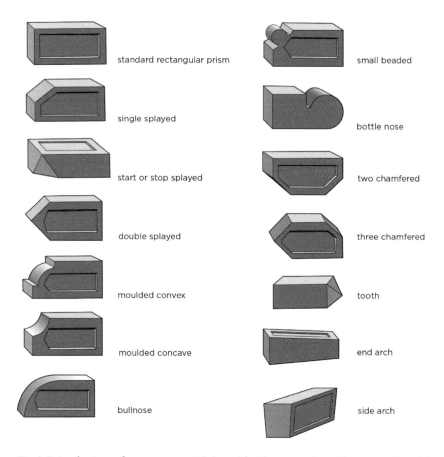

Fig. 3.7 A selection of common special shaped bricks; given the wide range of specials available in the 19th century, identifying the particular shape may not be possible, in which case the term "special moulding" can be used.

- *Bullnose bricks:* These have one corner rounded-off and are most commonly found on the corners, or "quoins", of buildings where a sharp edge is not desired.
- *Coping bricks:* These were manufactured in a wide range of sizes and shapes with a sloped or pitched top, and were made to form protective copes on walls.
- *Arch bricks:* These were manufactured to be used to form arches. Correctly termed voussoirs, they were made in the shape of a truncated wedge with the sharp end cut off.
- *Circle or compass bricks:* Used to build curved elements within a building which had a tight radius and other curved structures such as industrial chimneys.
- *Splayed or cant bricks:* Bricks with one or two corners cut diagonally; these were generally used at corners where a sharp edge was undesirable.
- *Air bricks:* Manufactured with perforations to allow the passage of air, these bricks were incorporated where ventilation through the wall was required.

Colour

The colour of bricks comes from a combination of the mineralogical content of the raw materials employed in their manufacture, the temperature reached during firing and the level of oxygen in the kiln. The colour of bricks varies considerably; most commonly, red and orange colours dominate brickwork, but white, cream, yellow and buff bricks are also frequently found. These were often employed in polychromatic brickwork, as considered in section 4.3. Darker bricks closer to blue or black in colour may also be found.

There is, at present, no easy-to-use colour chart to reference the colour of traditional bricks. However, it can be useful to use resources such as the Munsell Soil Colour Chart or the Munsell Rock Colour Chart when categorising the colour of bricks. Glazed or enamelled bricks can present a very wide range of colours, differing markedly from those that could be considered as naturally occurring in brickwork, due to the glaze applied during the manufacturing process.

Brick colour can be an important part of assessment, as identifying the colour of bricks can inform the sourcing of replacements in future repair and may have an influence on the availability and cost of materials should any replacement bricks be required. It can also help differentiate between one area of brickwork and another, this possibly being indicative of different phases of construction. The colour of a brick will be of particular significance where a polychromatic decorative scheme is present or where glazed bricks are used, as it will form an integral part of the aesthetic of the building. Where polychromatic decorative schemes are present within a building, these should be recorded as accurately as possible, as noted in section 4.3.

Fig. 3.8 Where bricks of the wrong colour are used in repair work, the impact on the aesthetic of the building is considerable.

Fig. 3.9 A rare example of the use of both darker and lighter brick in polychromatic decoration; the colour of the brickwork is clearly of considerable importance.

Surface texture

The surface texture of brick is influenced by various factors, including the raw materials used and the moulding technique employed. Where bricks were manufactured using sand moulding, for example, they will generally have characteristic crease marks, sometimes referred to as "the brickmaker's smile". This occurred due to the way clay was cast into the mould box. The surface texture of handmade bricks can also be influenced by the coarseness or otherwise of the sand used to dust the mould and table. Where bricks were manufactured by extrusion, drag marks will often be found on the surface of the brick, where small inclusions in the clay have been dragged across the face of the bricks. Those manufactured by mechanised pressing are more likely to exhibit a smooth, uniform texture. Gauged bricks will also generally have a very smooth, rubbed texture due both to the method and materials used in manufacturing and the rubbing process to form gauged work discussed in section 4.2. Bricks manufactured from shale are likely to have what may be termed a "granular" surface texture. The surface texture of bricks is a further feature to note in survey work.

In some cases, it may be found that the name of a brickmaking company has been pressed into one of the bed faces of the brick. This was carried out where bricks were manufactured by mechanical methods. In cases where pressed bricks were manufactured, this was executed during the pressing

Fig. 3.10 Some 18th century handmade bricks laid in irregular bond.

process. Where bricks were formed by extrusion, these too would some-times have the name of the brickmaker lightly pressed onto the face of the brick manually, using a stamp. Handmade bricks may, on occasion, also be marked with lettering or other marks inscribed into the wet clay. Where this is found, its significance in terms of heritage, and potentially archaeology, should be noted.

Depending on the arrangement of bricks in a kiln, what are termed "kiss marks" may be found on the face of a brick. This related to the disposition of bricks within a kiln with areas overlapped by a brick above having less direct exposure to firing, leaving a characteristic mark as shown in Figure 3.11. Kiss markings are not indicative of deficiencies with the bricks, but can affect the aesthetic of a building.

The shape, colour and surface texture of bricks can be seen to be important characteristics of traditional brickwork that need to be carefully assessed and recorded during survey work. All contribute to both the aesthetic and technical performance of a brick structure, and are important factors in any repair work that may be undertaken following a survey.

Fig. 3.11 Variegated appearance and "kiss marks" on colliery shale common bricks.

Fig. 3.12 This brick, dating to the early 20th century, has an applied surface texture; this may have been designed to act as a key for render or may simply have been a decorative embellishment.

3.3 Gauge

When bricks are laid in a wall, the mortar joint between each brick horizontally is known as the bed joint. One horizontal row of bricks is known as a course. The height of the bed joint has a considerable influence on the appearance of a brick wall. The most common way, historically, of expressing the height of a bed joint in brickwork was not to consider each individual course, but rather the height to which a given number of courses should rise – most commonly four courses of bricks with the bed joints between. This is known as the gauge of the brickwork.

A number of historic reference works can be used to illustrate the use of gauge in defining brickwork. One 18th century work specified that "four courses of brick should not rise above 11 inches" (Surveyor 1787, p. 2). Pasley (1826, p. 232) notes that four courses of brickwork will average 12 inches in height "in rough work". A number of technical reference works provide instruction on the height to which four courses of bricks should be built in the 19th century. Christy (1882, p. 17) describes a joint of above a quarter inch as "coarse work", giving a height of 12 inches for four courses of brickwork. Hasluck (1905, p. 21) states that four courses of bricks should equal one foot in height.

As with a number of other technical features of traditional brickwork, there is considerable regional variation in gauge. In the North of England "Northern Gauge" brickwork is recorded, as described by Lynch (1994, p. 14), who notes that brickwork in the North of England would have four vertical courses measuring 13.5 inches (34 cm); to the South this would be only

12 inches (30.5 cm). In Scotland, survey of many traditional brick buildings has shown that the 19th century saw the development of a specific Scottish gauge, where four courses of brickwork would rise to 14 or 15 inches (25.5–38 cm) in height. This again demonstrates the importance of developing and utilising knowledge of regional and national craft practices when surveying traditional brick buildings.

It should be noted that buildings may utilise varying gauges of brickwork in different parts of the structure – for example, where a front façade of brickwork is constructed using pressed facing bricks, and a side elevation is constructed using common bricks, these may well be built to different gauges. Likewise, where facing bricks are backed up using common bricks, or where glazed bricks are used as a facing material, there is a likelihood that the gauge of the facing bricks and the backing brick will be different (although best practice would be for the two to be built to the same gauge to ensure strong bonding). Where different gauges are used in a building, and in particular where they are used in the construction of a single wall, this can cause difficulties in properly bonding the areas of brickwork together. All the gauges used in a building should be noted in survey work and any possibility of these causing defects or instability also highlighted. Gauge is an important aspect of brick construction,

Fig. 3.13 Differences in gauge can clearly be seen between the original brickwork on the right and the modern repair on the left.

as it informs repair work and rebuilding, and highlights potential instability between walls. It can also be an indicator of different phases of construction or previous repair work.

When measuring the gauge of a piece of brickwork, this measurement should be taken from the upper arris of one brick to the lower arris of a brick three courses below. The gauge measurement therefore includes four bricks and three bed joints. A gauge or gauging rod was the tool traditionally used to check the accuracy of the gauge in brickwork. This can be a helpful tool when surveying brickwork as it provides a quick reference point to compare the gauge in one area of brickwork to another.

3.4 Joints and joint profiles

When bricks are laid in mortar to construct brickwork, mortar joints are formed where a brick meets its neighbour. The joints that run vertically between bricks in a course are often referred to as perpend joints or shortened to simply "perps". The joints that run horizontally between courses are often referred to as "bed joints", as these are the joints onto which the course above is bedded. The joints between bricks in a wall one brick thick or greater are generally referred to as "collar joints". When visually inspecting mortar joints in traditionally constructed brickwork, there are several distinct aspects to consider, including the condition of the mortar, the height and width of the joints, the materials used and any evidence that may be apparent regarding composition, joint finishes and method of use.

When surveying brickwork, it will often be appropriate to measure the size of both the bed joints and the perpend joints. The width and height of mortar joints are an integral part of a brick structure. It is good practice to record a variety of different joint heights and widths across an area of brickwork being surveyed, as this will allow for any variations to be accounted for. Joint height and width are related to gauge, as discussed above. With regards to collar joints, the most common requirement when surveying is to ensure that these have been well filled during both construction and later repair. Unfilled or partially filled joints of any type are a source of vulnerability for brickwork and can contribute to the decay of bricks and brickwork. Defects in mortar joints are discussed in section 6.5.

A further point to note regarding mortar joints is the type of mortar present. At its most binary distinction this is often defined as either "lime" mortar or "cement" mortar, which is an over-simplification, as discussed in Chapter 2. The visual identification of mortar type can be hard, and differentiating between different types of lime mortar, cement-based mortar and earth mortar is difficult to do by eye only – even for those with experience. However, it will often be the case that cement-based mortars are causing damage to the bricks. In such cases, visual identification of this may be possible and should be recorded during survey. When seeking to identify the type of binder used or the ratio of binder to aggregate present within a brick wall, it is likely that recourse will have to be made to mortar analysis, as discussed in section 7.1.

A wide range of finishes could be applied to mortar joints. These are summarised in Figure 3.14. A distinction should be made at this point between "jointing" and "pointing" of brickwork. Jointing is the practice of finishing the bedding mortar of brickwork with a joint profile. Pointing is the practice of raking out some of the bedding mortar and pointing this joint with a higher quality mortar. A joint finish would then be applied to the pointing mortar. Where evidence of joint finishes is found, this should be highlighted in survey work.

Joint finishes take a number of forms. When original joint finishes are found during the survey of a brick building, it is important that these are recorded accurately. This is likely to involve both a written and photographic record. It is often the case that joint profiles have largely been lost either as a result of the natural wear and tear of a building, or through previous inappropriate repairs. It may be that only a small section of original joint profile survives in a relatively sheltered part of a structure. Where this is seen to be the case, again the joint profile should be recorded carefully as it may be that this joint profile is replicated in future repointing work. Case study 2 is a good example of where limited evidence of surviving joint profiles informed later repair work. Some joint profiles, such as tuck pointing, are particularly vulnerable to decay, and where these survive it is likely that only small, isolated patches will be visible to the surveyor.

Other characteristics that can be recorded regarding mortar include colour and exposed aggregate. The colour of mortar is influenced by both the binder and aggregate used. Mortar could also have pigments added to bring about a change of colour for aesthetic reasons. This was most commonly carried out to blend mortar joints with the colour of surrounding bricks, as seen in Figure 3.15. If exposed aggregate is visible in mortar, this should be noted during survey and assessment. In some types of mortar joint – for example, the very fine joints between bricks in gauged work, aggregate is likely to be so fine that it will not be visible. If a mortar has been formed as a "hot mix", there may be particles of unslaked lime visible in the joints, which will show as white particles in the mortar joints.

Differentiating between original mortar and later repairs can be difficult using only visual inspection. This also raises the question of what should be considered "original" in a conservation sense as, in brickwork of considerable age, it is likely that this has been repointed many times. Experience plays a large part in this process. Despite the use of early cement, where a mortar that uses cement as a binder has been used, this will often be a later intervention. If this is found to be the case ,the careful cutting out of the later mortar may reveal earlier mortar behind it, which can then be used as a basis for repair work. There is also the potential to find original sections of mortar that survive in less-exposed parts of a building – for example under eaves (Figure 3.15).

In general, recording the condition of the mortar joints and any of the visual characteristics described above will form the extent of what can be achieved in a general survey. If more detailed information about the composition of mortar is required, recourse is likely to be required to mortar analysis as discussed in section 7.1.

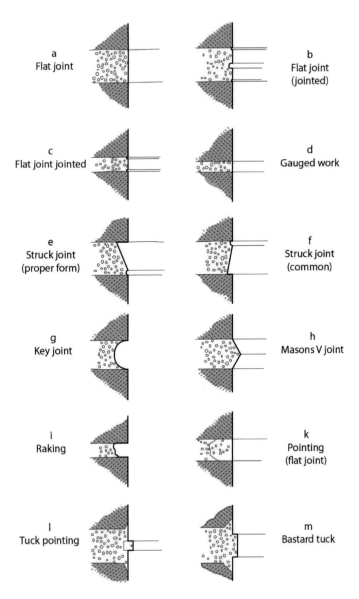

Fig. 3.14 Joint profiles commonly found in traditional brickwork (based on a drawing from Pasley 1826).

It should be noted that there are limits to what both visual survey and scientific analysis can provide in terms of specification for repair mortars. A wider understanding of the properties of the bricks used, the exposure of the brickwork and other factors that may affect the mortar joints is required to create an appropriate specification for replacement mortar. For example, brickwork forming an exposed chimney stack is likely to need a more durable mortar than that for a sheltered wall in an urban environment. Gauged brickwork will require a different strength of mortar with a different aggregate than brickwork formed of colliery shale bricks. When surveying brickwork ahead of repair work that requires repointing or bedding of new bricks, working with a specialist in lime mortars may be required in addition to the visual survey and analysis discussed in this section.

Fig. 3.15 A small patch of a tuck-pointing style joint finish found surviving on a 19th century building; this only survived where the brickwork was protected from the elements under eaves. Evidence such as this should be recorded carefully.

Fig. 3.16 A surviving section of ruled joint finish in an 18th century military structure.

Fig. 3.17 The formation of a ruled joint during a trial ahead of repair work.

3.5 Arches

In any brick structure, there will be a need to create openings for windows, doors and numerous other purposes. In traditionally constructed brickwork, such openings were either bridged by a lintel or an arch. In areas where stone was relatively widely available, stone lintels were sometimes employed above openings, generally with a brick relieving arch above. However, in many traditional brick buildings arches were the means by which openings were bridged. In common with the majority of aspects of traditional brickwork, arches have a diverse range of features. There are several characteristics of arches in traditional brickwork that require definition during survey work. These include form, depth, bond and method of construction (often referred to as the type of arch). In some forms of arch, the angle of skewback can also be noted.

If, firstly, the method of construction is considered, arches are generally defined as being constructed as "rough", "gauged" or "axed" arches. The type of arch can be defined in terms of the method used to shape the individual voussoirs. This differentiation is also defined as either being "segmental" or "rough", segmental arches being those in which the bricks are cut or manufactured so that they taper to the radius of the arch with parallel joints in between and rough arches being those where the bricks are un-cut and the joints do not taper. This is a somewhat confusing method of differentiating between arch construction, though, as the term "segmental arch" is also used for arches where the curve is less than a semi-circle, as discussed below. For this reason, the terms "axed", "rough" and "gauged" are generally more appropriate when surveying and recording traditional brickwork.

Where a rough arch is constructed, the bricks are not cut, but are built into the arch in the rectangular prism shape to which they were moulded. This leaves a tapering or V-shaped joint between each individual brick voussoir in the arch. Where an axed arch is constructed, the bricks are cut to form the arch voussoirs. Not as accurate as a gauged arch, an axed arch nonetheless presents a more stable form than a rough arch. The name "axed arch" comes from the use of the brick axe to cut the bricks to shape. The highest expression of brickwork in terms of arch construction comes with the building of a gauged arch. Using techniques noted in section 4.2, where a gauged arch is constructed, each individual brick voussoir is accurately cut and rubbed to allow for a joint of less than 2 mm between each brick. Arches were increasingly supplied in pre-moulded sets of special bricks from the 1860s onwards. These special bricks were often numbered at the time of manufacture and supplied for assembly to construction sites. This was particularly common where polychromatic arches were being formed, with brickworks supplying arches to various dimensions as required. The method of construction of an arch should always be noted in survey work to inform any future repair, maintenance or conservation work.

The form of an arch in brickwork also requires to be defined in survey and assessment work. This is defined by the shape of the arch with segmental, camber and semi-circular being the three most common forms used in traditional brickwork. In simple terms, a segmental arch is one with an extrados (the outer curve) and intrados (the inner curve) formed of a curve less than a semi-circle. Segmental arches can additionally be described by their depth, bond, type and angle of skewback. A semi-circular arch, as the name suggests, is one that takes the form of a semi-circle. Technical authors widely regard it as the strongest form of arch. Semi-circular arches spring from a horizontal bed. They can further be described by their depth, bond and type. A camber arch can be defined as one which has a horizontal extrados and only a slightly curved, or "cambered", intrados. The rise of the curve on a camber arch is often fairly slight, with Gourlay (1903, p. 9) noting this as being roughly an eighth of an inch per foot of span. The joints between bricks in a camber arch must radiate to a supposed point much closer to the centre than would be the case in segmental arches. A camber arch is differentiated from a straight arch due to the slight curve of the intrados.

A wide range of other forms of arch can be found in brickwork apart from the three principal forms described above. Also seen in brickwork are the Gothic arch, defined as "one whose curves have each an angle of an equilateral triangle as its centre, the two meeting at the third angle", and the elliptical arch, defined as one that "has one half of an ellipse for its curve ... used for wide spans where only a small rise is obtainable". As the name suggests, a three-centred arch is a complex piece of construction, the arch being set out and constructed from three centres; in many ways it is akin to an elliptical arch. Where forms of arch that are uncommon are found in survey work, reference can be made to the works in the References and Further Reading section at the end of this chapter to aid in their identification. A round opening within brickwork can be formed in the formation of an oculus, often termed a bull's eye window. As with other

Fig. 3.18 The form of this arch is semi–circular: it is two bricks in depth formed of four concentric rings of header bricks.

Fig. 3.19 This oculus is a single brick in depth; the bond is alternating courses of a single stretcher and two headers. Decorative embellishment here includes the use of polychromatic brickwork and special bricks.

forms of arch, oculi can be defined in terms of depth and bond. They were generally found as part of wider decorative schemes and are therefore more likely to be used in polychromatic brickwork and to utilise special bricks.

The depth of an arch can also be defined; this is generally expressed as the number of bricks that form the distance between the extrados and intrados of the arch. A depth of a single brick is most commonly found but, as seen in Figure 4.22, arches of a depth exceeding two bricks is not uncommon. The depth of an arch can also be expressed as a measurement in millimetres

Fig. 3.20 The V-shaped joints between individual brick voussoir in this arch indicate that it is of rough construction. The arch is a segmental arch, a single brick in depth formed of a single header in every course; the angle of skewback is 78 degrees.

(or indeed feet or inches), so the arch in Figure 3.22 can be said to have a depth of one brick, or 230 mm in this case.

The bond of an arch, defined in terms of the relationship of headers and stretchers, can also be noted. For example, the arch in Figure 3.20 is one brick in depth, formed entirely of stretchers. An arch may also be formed of concentric rings of header bricks, with each header brick conforming to half a brick in depth. Various combinations of header and stretcher bricks can be found used in arches. This can see, for example, the use of alternate header and stretcher bricks in each course. The bond of an arch should be recorded as part of any survey of a traditional brick building.

Lastly, in arches that are not semi-circular in form, the angle of skewback can be measured. This is the angle at which an arch springs from the main body of a wall. An angle of skewback of between 70 and 80 degrees is commonly found but can be more acute than this as seen in Figure 3.21. The angle of skewback can be an important factor in survey ahead of repair work where arches are being taken down or other significant work is being undertaken.

As with other aspects of surveying traditional brickwork, with arch construction it is important to consider what lies beneath. Often a lintel will run across an opening in brickwork. This may be in addition to a structural arch, or it may be that the arch which shows on the external face of a brick wall is one or

Fig. 3.21 This camber arch has been constructed of special bricks. Each brick would have been numbered in order to make construction easier; the arch is a single brick in depth and has a very acute angle of skewback at 45 degrees.

even half a brick thick, and while fulfilling some structural requirements of the building, it is the lintel that takes the majority of the load. This is complicated further by the fact that above the lintel there is likely to be a brick relieving arch. It will therefore be seen that in some brickwork there may be an external arch showing on the face of a brick façade, a timber, stone or metal lintel behind and a relieving arch above this.

Fig. 3.22 A three-centred arch, a single brick in depth; arches of an uncommon form can require some research to ascertain their form.

Fig. 3.23 The intrados of this arch can be seen here with the wedge shape brick voussoirs also evident.

3.6 Reinforcement and damp-proof courses

As well as bricks and mortar, a number of other materials are commonly found as part of traditional brickwork. These materials are most commonly used to add reinforcement to brick structures or to form damp-proof courses close to ground level. Correctly identifying these materials can be an important part of survey and assessment work.

Brickwork as a construction method is generally strong in compression, in common with other types of masonry. However, in some structures, materials such as timber or iron were used to provide longitudinal reinforcement to brickwork. Where this is found to be the case during assessment of brickwork, it should be recorded. As such reinforcement is often hidden from view and does not show on the outer façade of brickwork, it is often only when it fails that it is noted.

Timber was built into brickwork for a number of purposes. Most commonly, this took the form of long pieces of timber built into brickwork known as chain timbers and bond timbers. Timber, which is completely encased within masonry so that no part of it is seen after the brickwork is completed, is known as chain timbers. The reason for using such timbers was to bond brickwork together and lessen the effects of settlement. Bond timbers differed from chain timbers in that the former were not entirely encased in masonry but had one face exposed on the internal face of the brickwork. This could be used to fix internal linings such as lath and plaster. Chain timbers were thought preferable to bond timbers as they were less likely to suffer from rot or decay, or be affected by fire.

In 1825, Marc Brunel devised a new system for reinforcing brickwork using thin strips of iron in what became known as "hoop iron bond". This was formed by inserting lengths of iron longitudinally in every fourth or fifth course of brickwork and had the benefit over bond or chain timbers of not posing the same level of risk in the event of fire or rot. It was noted that such hoop iron should be tarred and sanded, one row being used to every half brick in thickness and every 2 or 3 feet in height, hooked at all joints and angles. The iron used for such bonding was noted as "long thin strips of wrought iron 1.5 inches wide by one-sixteenth of an inch thick". Where there is evidence of any form of reinforcement in brickwork, this should be noted and a careful assessment of defects related to the failure of such systems carried out.

A damp-proof course (often abbreviated to DPC) is a continuous layer of material incorporated into brickwork close to ground level to prevent the upward movement of moisture from the ground. It is unclear exactly when damp-proof courses began to be incorporated into brickwork. Brunskill (2009, p. 241), in his summary of the history of DPCs, indicates that the mid-19th century marked the beginning of their use with technical reference works and survey work supporting this. A number of materials have been used to form damp-proof courses in traditional brickwork, including bitumen, slate and lead.

Where used in traditional brick buildings, these can generally be seen between courses of brickwork close to ground level. The efficacy of these damp-proof courses varies considerably and is influenced by the way they were laid originally, modern ground conditions and any actions taken that may have bridged the damp-proof course.

Damp-proof courses have also come to be installed in brickwork as a retro-fitted measure. The use of more modern injected damp-proof courses can generally be identified through a series of holes drilled through bricks and mortar close to ground level. These will often have been plugged using a cement-based material. Injected damp-proof courses of this type, if installed inappropriately, can cause damage to adjacent brickwork. The presence of a damp-proof course,

Fig. 3.24 Two early forms of damp-proof courses can be seen in this 19th century structure, a layer of bitumen and a layer of slate.

Fig 3.25 A thick course of slate has been laid as a damp-proof course close to ground level.

Fig. 3.26 A modern injected damp-proof course can be seen here through the small, plugged holes close to ground level.

either original or modern, should be recorded in survey work. The majority of traditional brick buildings would have been constructed without a damp-proof course and its absence should not be interpreted as a deficiency in a building.

3.7 Methods of constructing brickwork

When surveying brickwork, one of the fundamental technical characteristics to consider is how the walls have been constructed. There are broadly three methods of constructing a masonry wall with brick commonly found in the United Kingdom: solid brick walls; cavity brick walls; and composite walls of brick and a second material. Each of these methods will generally present those who have a need to assess brickwork with indicators of which has been used. As with all traditional construction, however, the unexpected is always a possibility and gaining as much information as possible about a structure will help with the identification of which form of construction is used. The following sections aid in such identification but are by no means exhaustive, as the variations in the ways brick has been used throughout the world are almost limitless.

3.7.1 Solid brick walls

The majority of traditionally constructed brickwork falls into the category of what may be termed solid wall construction. Walls constructed of solid brick-work are the form most likely to be encountered when working with tradition-ally constructed buildings. A solid brick wall is one in which the brickwork acts as the principal structural element of the wall, the bricks generally being laid in a bond pattern to ensure strength and stability (although they may, in some cases, be laid in irregular bond). Walls of solid brickwork are found in Britain, and indeed worldwide, from the earliest periods of use of the material.

The thickness of a solid masonry wall of brickwork will vary depending on various factors, including the height of the structure, the period in which it was built and the local climatic conditions. Wall thickness is generally defined in terms of how many brick lengths the wall is. Therefore, a wall that is the length of one brick in thickness is defined as a wall one brick thick. Walls one and a half bricks in thickness are common in domestic structures in the United Kingdom, such as terraced houses. Engineered structures or large industrial buildings will have walls of greater thickness, however. As with gauge, the thickness of a solid brick wall can also be expressed as a numerical distance in millimetres so, for example, the wall in Figure 3.26 can be noted as being one and half bricks in thickness, or 343 mm.

When building in solid brickwork, bricks will be laid as headers and stretchers in a bond pattern, as described in section 3.1. These bond patterns utilise headers and stretchers for both structural integrity and aesthetic qualities. This bond pattern is one of the clearest indicators that a wall is solid brickwork, although as will be seen, this is not an infallible rule to follow.

3.7.2 Cavity walls

A cavity wall is one which purposefully contains a void or cavity between two masonry walls. These two walls are generally referred to as either "skins" or "leafs" and can be formed of brick or stone masonry. The use of cavity wall brickwork has a longer history than may be expected. Early prototypes of brickwork that incorporated a cavity between two brick walls began to be used in the United Kingdom in the 1820s. Indeed, a prototype of cavity walling was first conceived of by an English architect, William Atkinson, as early as 1805 (Atkinson 1805, pp. 15–16), with Thomas Dearne also developing an early form of the technique. Early examples of this form of construction utilise brick to tie across the cavity in various bond patterns such as rat-trap bond, which also utilised brick laid on edge.

Most early forms of cavity walling utilised header bricks to tie across the cavity. Later, cast iron wall ties began to be used to link cavity walls together. The identification of cavity wall when surveying brickwork presents some challenges, depending on the method of construction used. Where metal wall ties are used to tie the two leaves of brickwork together, this is most commonly identified by the use of stretcher bond. However, where bricks are used to tie across the cavity, this is more challenging to identify as a more recognisable bond pattern, similar in form to that used for solid brick walls may present on the front façade. In some cases, special bricks that were more highly vitrified were used to tie across the cavity, as seen in fig. 3.29. A further complication comes when it is considered that in the 19th century the outer layer of brickwork may have been built as a solid wall one brick thick as can be seen in Figure 6.14 in

Fig. 3.27 This building utilises brick on edge to form what is termed "rat–trap bond"; this forms a honeycomb cavity between the outer and inner brickwork.

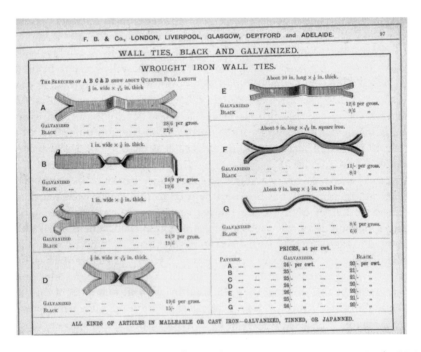

Fig. 3.28 A selection of cavity wall ties. These came into common use in the United Kingdom from the latter half of the 19th century and were available in many patterns.

Fig. 3.29 The header bricks interspersed in this wall tie across a cavity. They are denser and more vitrified than the common brick used in the majority of the wall. These bricks were made larger to bridge the cavity.

Chapter 6. Where situations such as this are encountered, it can be hard to identify the wall as being of cavity construction. Where doubt exists as to whether a wall incorporates a cavity, the specialist investigative techniques discussed in Chapter 7 may be required to ascertain the method of construction.

In some cases, cavity walls were formed using bricks laid on edge, as seen in what is termed "rat-trap bond" illustrated in Figure 3.27. Where bricks are laid on edge in this fashion the cavity is frequently bridged by bricks tying across. This can be hard to identify in survey work, but careful measurement of bricks should reveal where they are laid on edge.

3.7.3 Composite walls

The third method of constructing a wall using brick is to employ it as part of a composite construction with another material. There are many different combinations of materials that use brickwork in part of their construction; this section discusses the identification of three of them: brick-lined ashlar, brick lined rubble and brick/clay construction. This is an indicative rather than exhaustive discussion, however, and an area of construction where considerable regional and national variation may be found.

Brick-lined ashlar, sometimes referred to simply as brick ashlar, is a form of construction whereby the front façade of a wall formed of ashlar stone masonry has a backing of brickwork. The ashlar will generally be coursed to block bond with the brickwork. This form of construction was relatively common in the 18th and 19th centuries and, towards the end of the 19th century, could include a cavity between the ashlar outer leaf and the brick inner leaf. As it is the inside face of a wall that is formed of brickwork, brick lined ashlar can be a hard technique to identify unless sections of internal linings such as lath and plaster are opened up – a disruptive and invasive process. Where brick is used to line ashlar walls, this should be recorded as there are potential issues around the two materials suffering from a lack of bonding; where a cavity is incorporated, the potential defects common to all cavity walls, as discussed in section 6.4.3, may be found.

Brick was also used to form composite walls with rubble stone masonry. In good practice, the rubble work could be coursed to bond with the brick backing, although this was by no means ubiquitous. It is most commonly seen in the construction of garden walls, which were lined on their inner face with brick due to the material's superior heat retention and consequent beneficial effect on the cultivation of plants. Where such walls were constructed well, header bricks would tie the brick lining with the rubble walling. However, this was not always well executed and it can lead to problems of delamination of the brick and rubble parts of the wall. This is discussed more fully in section 6.7.

Into this consideration of hybrid construction would also fall the myriad uses of brick in conjunction with walls formed of stone masonry – for example, in the forming of arches, quoins, flues and reveals, all of which may be found constructed of brick in traditional buildings where the principal construction material is stone. This is generally found where the stone is of a type that is not easily dressed to a flat surface, making the use of brick at reveals and to form arches a necessity. It is also sometimes found that brick-built flues were constructed in buildings that utilise stone as the main masonry material due to superior resistance to the actions of heat. Brick is also found in civil engineering structures such as bridges, which may be constructed largely of stone but with the load-bearing arch formed of brick. In particular, this is found in skew arch bridges, which were easier to set out and construct in brick than stone.

For the same reason, brick can also be found used to construct vaulting in buildings where stone is the main building material. Where brick is found as a material put to a particular purpose in a largely stone build structure, the same considerations in terms of technical, aesthetic and decay characteristics that apply to traditional brickwork used as a principal masonry material should be considered.

In vernacular construction, a wide and disparate range of hybrid clay and brick construction techniques may be found. They include over-cladding, permanent shuttering and the use of clay-based mortar. In areas where rammed earth construction was used, brick has been found used as a permanent shuttering on both the internal and external face of the earth core. The brickwork was generally formed of two half brick skins laid in stretcher bond with headers placed intermittently to bond into the earth core. It is noted by one technical author that ties similar to cavity walling were sometimes used to aid the bonding of the brickwork with the earth material (McGregor 1996, p. 54).

A further use of brick in conjunction with earth construction is the over-cladding of clay or earth walled buildings with brickwork as a retrospective measure. The brick over cladding would generally be formed of either half a brick-thick walls laid in stretcher bond or as brick on edge, both of which saw headers used to bond the brickwork to the clay wall behind and closers inserted where required to ensure that an irregular bond is maintained. In the example seen in Figure 3.30B, this is done successfully with an almost total lack of straight vertical joints between courses throughout the build. When the over cladding was being applied, the clay wall would have been cut back to allow the brickwork to sit on the original rubble foundation, with rebates being cut to allow the headers to bond into the clay wall (Copp 2009, p. 60).

Clay and earth-based mortars have a long tradition of use in some parts of the United Kingdom. Although it is more commonly found used in conjunction with stone, it is possible to find brickwork laid in a clay-based mortar, especially where there is a tradition of this in a local area. This may have been repointed with lime or cement-based mortars in later years. Where an earth-based mortar is found during survey of brickwork, this should be recorded as being of significance.

A further wall type that can be regarded as composite walling is the use of brick to form infill panels between structural timber framing. In the United Kingdom, this can be found dating to as early as the 15th century. It is found as an external wall construction in England and Wales. In Scotland, however, it is only found in the construction of internal walls. This further demonstrates the importance of understanding regional variations in brick construction. Such infill involves the construction of a structural timber frame with brickwork being built within the frame. The distance between the horizontal and vertical timbers varies in such work, with one 18th century technical author noting it as being around 3 feet apart vertically (Surveyor 1787, p. 9). In well-executed infill work, the timbers would be set at measured distances apart to allow for a given number of courses of brickwork and mortar joints, although this was not always the case. In some cases, and particularly where the infill work was forming an internal partition, the brick would be laid on edge.

When surveying brick infill, the same considerations regarding deterioration of bricks and mortar apply as for other forms of brick construction.

With brick infill, however, the condition of the supporting timber frame will clearly be of considerable importance. There may be disruption or alteration to bond patterns in infill work to facilitate the timber frame. Decorative bond patterns, such as herringbone or basket weave, may also be encountered with brick infill.

With traditional brickwork, appearances can sometimes be deceptive. Composite walls could also be formed of different types of bricks. From the

Fig. 3.30 a and b The brickwork in this building has been formed of brick on edge. It has been used to over-clad a building originally of earth construction. In fig 3.30a, the earth is evident behind the brickwork.

Fig. 3.30b

Fig. 3.31 An internal wall constructed of brick. The outside façade of this building was stone, but the internal partitions along with the side and rear elevations were constructed from brick. The internal wall is only half a brick thick and is constructed of common brick laid in thick joints, a common practice for internal walls.

Fig. 3.32 Although the majority of this building is constructed of rubble stone, the arch above the window is formed of brick. The building dates to the mid-18th century and the brick arch is original to this date.

Fig. 3.33 Brick infill panels are used at the gable of this building. The vertical and horizontal timbers can clearly be seen.

Fig. 3.34 The brickwork in this image is formed of brick infill panels between timber. This is an internal partition, but in some parts of the United Kingdom such infill work was used externally.

17th century onwards, there was a practice whereby a brick wall would have a half brick-thick façade of facing bricks with a backing of cheaper brickwork. To further economise, in some buildings only a small number of headers tied this façade of facing brick into the backing, the majority of headers being "snapped", meaning they did not tie into the brick backing. This poor bonding can result in instability of a wall with the front façade delaminating from the backing. This will be apparent in survey work by bulging or leaning of walls and also by cracks emerging in the brickwork. Simply because a wall seems to be well bonded, that does not necessarily mean it is. Poor bonding, therefore, is something to be mindful of where structural issues are apparent. Where a higher quality of brick was used on the front face of a wall, it is often the case that the bricks behind are of a poorer quality. As such, if these bricks have become exposed, they are likely to suffer from decay and deteriorate at a faster rate than those on the front façade.

3.7.4 Internal walls

Internal walls in traditional buildings are often found constructed of brick, even where stone is used for external walls. As one technical author notes, "even in the heart of stone districts bricks are now generally used for all internal walls ... Brick walls are also straighter than rubble walls and require less plaster." (Sutcliffe 1898, p. 91). These advantages of brick over rubble are certainly true, and it may be added that it is easier to build a relatively thin wall of brick than rubble.

The main survey requirements for internal brick walls are largely the same as for external walls. It should be noted that the characteristics of the bricks used for internal walls are likely to be different from those used externally. Internal walls were often built of bricks fired at lower temperatures or those in some other way rejected for use externally as they were not expected to be exposed to the elements. Internal walls are often likely to be thinner, in many cases half a brick thick. They may also have wider joints, as sometimes internal walls were built at a faster rate than external walls and were built in a rougher fashion because were never meant to be seen, as seen in Figure 3.31.

In some buildings, an internal wall may become an external wall over time. This may have happened due to demolition of part of a building and can be discerned where a brick wall has fireplaces or other internal features on its face, as shown in Figure 1.2. Such a wall may perform poorly in terms of durability, resistance to rain penetration and thermal performance. Where such a situation is found during survey work, it should be highlighted.

The decay mechanisms that affect internal walls are subtly different from those affecting external walls. The source of moisture affecting bricks internally will often be the result of condensation, both surface and interstitial. Faulty services can also be a source of moisture affecting internal walls. Problems can be encountered when surveying internal brickwork due to access, as very often internal walls will be finished with lath and plaster, plaster on the hard or later plaster board. If decay or structural instability is suspected in an internal brick wall, it may be necessary to open up or remove parts of these wall linings to allow for further investigation. This is an invasive procedure, however, and consent and permissions may need to be sought, especially for protected buildings or those of particular heritage significance.

3.7.5 Angled and curved brickwork

Where two walls form the corner of a structure, it is necessary to build brick-work at an angle to form what is termed the return of the wall. The corner of a building is also referred to as the quoin in traditional brickwork termin-ology. By far the most common angle at which brickwork is built is a simple 90-degree right angle. This is a relatively simple piece of brickwork to exe-cute; for example, where English bond is used, at a right angle a stretching course becomes a heading course in turning the corner and continues on as such. The creation of walls in brickwork becomes more problematic where they meet at an angle that is either greater or less than 90 degrees, as seen in Figure 3.36. This will require either bricks being cut to form the angle or the

Fig. 3.35 A simple right-angled quoin such as this is the most common found when surveying traditional brickwork. Any details such as the use of special bricks or polychromatic brickwork should be noted.

use of special bricks. Where brickwork constructed at an angle is found in the survey of a building, this should be recorded and the method of forming the angle noted.

In some structures, brickwork was also constructed to form a curve. For some work of this type, special bricks were used. If special bricks were not used, the use of header bond is sometimes found in curved brickwork. If the individual bricks were not cut to taper with the curve of the wall, the joints will be V shaped, increasing the mortar-to-brick ratio in the wall. The details of curved brickwork should be recorded in survey and assessment, with the additional consideration of how the curve is achieved.

Fig. 3.36 This obtuse quoin is formed of dogleg bricks. In the period when special bricks were available, this was a common method of constructing such brick-work.

Fig. 3.37 This building is almost triangular in plan; special bricks are used at the quoins.

Fig. 3.38 Curved brickwork could be formed of special bricks, or of bricks laid with V-shaped internal joints. Curved brickwork is often seen during survey and assessment.

3.7.6 Pilasters

A pilaster is a flat column that projects from the main body of a brick wall. They are used for several reasons in traditional brickwork, both for decorative and technical purposes. Where found in traditional brickwork, there are a number of technical features that need to be recorded in survey work. Pilasters can be defined by their width, depth and bond. Width is generally expressed in terms of numbers of bricks. The example shown in Figure 3.39 would be defined as four bricks in width although in most detailed surveys a linear measurement would also be taken, which in this case was 955 mm. The same is true of the depth of the pilaster, this referring to the projection of the pilaster from the face of the wall. In most cases, this will either be a half brick projection or a single brick projection but again a measurement is also taken. The bond of pilasters, as with all brickwork, relates to the disposition of headers and stretchers in order

Fig. 3.39 This pilaster projects half a brick from the main façade; it is four bricks in width and incorporates a polychromatic decorative scheme. The bonding is complex with courses as follows: stretcher/closer/header/header/header/closer/stretcher; header/stretcher/stretcher/stretcher/header; stretcher/stretcher/stretcher.

to avoid straight vertical joints in successive courses. This is made somewhat more complicated as the pilaster also has to bond into the main body of the brickwork, in some cases resulting in rather convoluted bonding arrangements, as in Figure 3.39.

3.7.7 Brick chimneys and flues

Brick was commonly used to construct chimney stacks and flues. This is seen in buildings where the main masonry material is brick; however, brick chimneys and flues can also be found in buildings where the main structural masonry is formed of stone or earth. Brick chimney stacks can experience significant problems and defects due to their exposed location and, for this reason, often require particular care during survey work. Chimneys are exposed parts of a building and are often constructed of relatively thin brickwork, in many cases only half a brick thick. In addition to this, they are exposed to the actions of salts and other contaminants from the flue gases produced by the fire that they support. Care should be taken in survey to note the presence of brick chimney stacks and flues movement, as well as their condition – for example, cracking or deformation of chimney stacks or chimney stacks that are out of plumb. Access to chimney stacks can be problematic, and viewing from ground level may not provide enough detail to fully assess the condition of chimney stack brickwork. Options discussed in section 7.7 may be necessary when surveying brick-built chimneys.

3.7.8 Vaulting and arched flooring

Brickwork has been used to form vaulting using various methods in the period that may be defined as traditional construction. A groined vault is one formed by the intersection of two or more cylindrical surfaces. Groined vaults were considered one of the most complex pieces of work for a bricklayer to

Fig. 3.40 A simple pilaster, projecting half a brick from the main face brickwork, this is one brick in width formed of alternating courses of headers and stretchers.

undertake due to the complexities of setting out and cutting the bricks. A simpler method of forming a vault is to construct a barrel vault. This takes the form of half a cylinder running longitudinally over the space being bridged or, to describe it another way, a semi-circular arch of elongated length.

By its nature, vaulting is often hidden from view. Where it is found in survey and assessment work, it should be noted carefully. The bond used in the vaulting can be recorded where a barrel vault is formed and also within the spandrel wall of groined vaulting. In terms of technical characteristics, accessible gauge and bond can be recorded as well as characteristics of the bricks themselves. Any signs of cracking or other structural defects, as covered in section 6.7, should be noted as vaulting will generally be supporting a considerable load.

Fig. 3.41 A brick-built flue in a rubble stone wall, a good example of the use of brick in a vernacular context.

Fig. 3.42 Brick is often used to form chimney stacks. This can be seen in buildings of brick and stone. As chimney stacks are exposed, they can be vulnerable to decay and are also hard to access.

Fig. 3.43 Vaulting such as this is a complex piece of brickwork. This Georgian era vault saw careful selection of bricks for use where cutting was required.

Fig 3.44 A simpler method of forming a vault is a barrel vault. This example again dates to the Georgian period.

The 19th century saw the use of what were often termed fire-proof mills. These were generally built using a framework of iron with shallow brick arched floors, sometimes referred to as jack-arched floors. These were constructed to reduce the risk of fire. There were many developments throughout the 19th century in the construction of fireproof floors, detailed discussion of which can be found in Swailes (2009). The use of such shallow arched flooring is most common in industrial or commercial buildings, although an early prototype of this form of floor was recently found in a Scottish domestic building. Where survey and assessment work finds such flooring used, this should be noted. The bond of the arch may be visible, although the most common configurations were either header or stretcher bond. In the majority of examples, the arches are constructed half a brick in thickness, with all bricks being stretchers. It may also be possible to record details of the configuration of how the brick arch springs from the iron beam that it abuts. Defects are likely to involve deterioration of either bricks or mortar. The example in Figure 3.46 is caused by abrasive cleaning seriously damaging the brickwork that forms the floor.

Fig. 3.45 Defects such as the patch of staining seen here are likely to require specialist investigation to ascertain their cause and to check whether the brickwork and supporting iron have become damaged.

Fig. 3.46 Shallow arch brick flooring, such as this supported on a framework of iron, was commonly seen in 19th century fireproof buildings.

Fig. 3.47 This is a rare example where glazed bricks have been used to form jack-arch flooring.

3.8 Case study 3: Hayford Mills, Stirling, United Kingdom – survey of technical features

The building that forms this case study exhibits many of the technical features of traditional brickwork set out in this chapter. The buildings that make up the mills were largely erected in the 1860s as a textile enterprise and have since been converted to housing. Two of the buildings and the tall chimneys that formed part of the complex have been demolished but the three that survive are excellent examples of mid-19th century industrial architecture.

The bricks themselves are of two types: the main facing brickwork is constructed of extruded wire cut bricks; the cream polychromatic brickwork is formed of a pressed facing brick. The size of the bricks in the main facing brickwork is length 225 mm, breadth 95 mm and height 85 mm. The polychromatic brick matches this at quoins and reveals. Scottish bond is used, with three courses of stretchers between each heading course. The gauge of the brickwork is four courses of brickwork rising to a height of 375 mm (equating to 14.75 inches). In addition to standard bricks, specials are used extensively for both decorative and technical purposes. These include arch sets, dogleg, single splayed, moulded convex, corner moulded convex and bullnose bricks. No curved brickwork is evident in the building, but in one section an obtuse angle is formed using dogleg bricks. This is a common application of special bricks in a practical rather than decorative context.

Three forms of arch are used in the structure: camber, semi-circular and three-centred. Camber and semi-circular arches are standard forms of arch in traditional brickwork, but the three-centred arch is much more unusual and complex in its geometry. The arches at the building were formed of specially moulded bricks produced by the firm Allan and Mann. This is known due to a neighbouring building being demolished and the rubble providing examples of the same arch as seen in the building surveyed. All the arches are a single brick in depth formed of single courses of stretchers; the angle of skewback for the camber arch is 78 degrees.

The pilasters, which are three bricks in width, employ a complex bonding arrangement. They are laid in Scottish bond, with three courses of stretchers between courses of headers (often abbreviated to 3/1) to match the main body of the brickwork. The three stretcher courses are formed of a course of two stretchers and two outer headers between an upper and a lower course of three stretchers. The heading courses are formed of two outer headers, two half brick closers and three inner headers in what is a sophisticated bonding arrangement.

It is worth noting in this consideration that the technical features of a brick building contribute to ascertaining significance. In this case, the building shows the earliest use of machine-made special bricks and also the earliest use of pressed facing bricks in Scotland. It is also one of the earliest uses of polychromatic brickwork, coming only eight years after the earliest example in Scotland. The 1860s and 1870s may be regarded as the golden age of traditional brick construction in Scotland, and this building was one of the earliest examples. Advances in the moulding and pressing of bricks contributed to this. The

Fig. 3.48 The extent of the decorative scheme used in the building can be seen here. It is clear that three forms of arch have been used, along with pilasters and other technical characteristics.

Fig. 3.49 The bond of the brickwork can be discerned here, both in the main façade and in the pilaster. The use of specials can also clearly be seen.

Fig. 3.50 This image demonstrates the complex interaction between aesthetic and technical characteristics, as the bond pattern to form the polychromatic decorative scheme has been disrupted.

technical features, therefore, add to the heritage significance of this building as well as providing a complete understanding of the structure itself.

There are clearly a range of decorative features as well as building defects, which an overall and comprehensive survey of the structure would record. This case study, however, has focused solely on recording the technical characteristics of the brickwork to illustrate the wider discussion in this chapter.

References and further reading

Adams H. (1906), *Building Construction,* London: Cassell.

Allen J. (1893), *Practical Building Construction*, London: Crosby Lockwood.

Atkinson W. (1805), *Views of Picturesque Cottages with Plans,* London: Gardiner.

Brick Development Association (2000), *BDA Guide to Successful Brickwork*, London: Butterworth-Heinemann.

Brunskill R.W. (2009), *Brick and Clay Building in Britain*, New Haven, CT: Yale University Press.

Christy A. (1882) *A Practical Treatise on the Joints Made and Used By Builders*, London: Crosby Lockwood.

Copp S. (2009) "The Conservation of the Old Schoolhouse at Logie, Montrose", in Vernacular Buildings 32, Forfar: Scottish Vernacular Buildings Working Group.

Fairbairn W. (1854), *On the Application of Cast and Wrought Iron to Building Purposes*, London: Weale.

Firman R.J. (1994), "The Colour of Brick in Historic Brickwork", *British Brick Society Information*, vol. 61, pp. 3–9.

Firman R.J. and Firman P.E. (1967), "A Geological Approach to the Study of Medieval Bricks", *Mercian Geologist* vol. 2, no. 3, pp. 299–318.

Forsyth M. (ed.) (2007), *Structures and Construction in Historic Building Conservation*, Oxford: Blackwell.

Gourlay C. (1903) Elementary Building Construction and Drawing, Glasgow: Blackie.

Greenhalgh R. (1947), *Modern Building Construction*, London: New Era.

Grier W. (1852), The *Mechanics Calculator*, Glasgow: Blackie.

Hasluck P. (1905), *Practical Brickwork,* London: Cassell.

Johnston S. (1992), "Bonding Timbers in Old Brickwork", *Structural Survey*, vol. 10, no. 4, pp. 335–62.

Lynch G. (1994), *Brickwork: History, Technology and Practice*, London: Donhead.

Lynch G. (2007), *The History of Gauged Brickwork,* London: Elsevier.

McGregor C. (1996), *Earth Structures and Construction in Scotland*, Edinburgh: Historic Scotland.

Morton T. (2008), *Earth Masonry Design and Construction*, Bracknell: BRE.

Nicholson P. (1838), *Practical Masonry, Bricklaying and Plastering*, London: Kelly.

Surveyor (1787), The Builder's Price-Book, London: Taylor.

Sutcliffe G. (1898) *Modern House Construction Vol. 1*, Glasgow: Blackie.

Swailes T. (2009), *Guide for Practitioners 5: Scottish Iron Structures* Edinburgh: Historic Scotland.

Vicat L.J. (1837), *Mortars and Cements*, London: Weale

Watt D. (2011), *Surveying Historic Buildings*, London: Routledge.

Wight J. (1972), *Brick Building in England from the Middle Ages to 1550*, London: Baker.

Wermeil S. (1993), "The Development of Fireproof Construction in Great Britain and the United States in the Nineteenth Century", *Construction History*, vol. 9, pp. 3–26.

4 Decorative characteristics of traditional brickwork

Drawing a clear distinction between the technical and aesthetic characteristics of traditional brickwork is problematic as there is a significant degree of overlap between the two. Glazed bricks, for example, and those of special shape, can be part of both the technical and aesthetic characteristics of brickwork. It may have been possible to simply include all the characteristics of traditional brickwork under a broad definition of technical characteristics; however, it is important to recognise that brickwork also has considerable aesthetic qualities. This is often overlooked when brickwork is considered, as the material is considered utilitarian in nature. Highlighting the need to record and assess aesthetic as well as technical features is an important aspect of developing a rounded understanding of the material and its uses. To return to the principal of definition presented in Chapter 1, for the purpose of the following section, decorative features of brickwork in this context are considered to be those employed for aesthetic purposes, and without which the building could still function from a technical perspective.

4.1 Diaper and early patterned brickwork

The use of contrasting colours of brick to form decorative patterns in brickwork is one of the most common, and also one of the earliest, forms of decorative treatment applied to brickwork. The earliest form this took in the United Kingdom was the use of darker coloured bricks to create distinct patterns in brickwork. This is often termed "diaper" work, as a common decorative pattern was the creation of diamond shapes using intersecting diagonal lines. Other patterns can be found, including saltires, crosses and, in some cases, heart shapes. The bricks used to form these patterns were generally "flared" or vitrified header bricks. These bricks were formed when wood was used as a fuel source during firing, as it meant the ends of some bricks were overburnt and took on a darker, deep-blue colour, or a grey-blue finish. Such bricks are still manufactured today for use in restoration and repair work, as seen in Figure 4.1. These headers were then used to create patterns in brickwork.

DOI: 10.1201/9781003094166-5

When surveying a building that makes use of diaper pattern brickwork, this should be recorded carefully. In some cases, such decorative schemes may be incomplete. This may have been part of the original construction of the brickwork if the bricklayers did not have enough headers of the appropriate colour or it may be a result of later interventions and poorly executed repairs. Diaper patterned brickwork can be a useful indicator of the age of brickwork, but it should be noted that some Victorian brick buildings imitated diaper patterns in polychromatic brickwork, as discussed in section 4.3. A combination of correctly identifying the type of brick used and the pattern should give an indication of whether this is the case.

Paint finishes and colour washes were sometimes used to give the impression of diaper patterns. A close inspection of the brickwork will reveal whether or not this is the case. If this is found during the assessment of traditional brickwork, as with other original paint and colour wash finishes, this should be recorded, as it is of considerable heritage significance and will require particular care during any subsequent repair or maintenance work. Specialist advice may be required if such early decorative schemes are found, in order to ensure they are recorded appropriately.

Fig. 4.1 Diaper work patterns at Hampton Court Palace are indicative of early decorative schemes (copyright David Pickles/Historic England).

Fig. 4.2 The results of a modern firing using wood as a fuel source (copyright HG Mathews).

4.2 Gauged brickwork

Gauged brickwork is the highest expression of the bricklayer's art, in which the specially selected "rubbing" bricks, or "rubbers", are prepared, cut and rubbed to very accurate dimensions after being fired. They are then laid with very narrow and precise joints. Generally, to be classified as gauged brickwork, these joints need to be less than 2 mm in width. The cutting and rubbing of gauged brickwork to such a fine degree of accuracy required exceptional workmanship and skill: it is the work of a brick mason rather than a bricklayer. Gauged brickwork is found in various parts of the world, but is most common in England, where it was used from the early 17th century, and although its use had declined by the early 1950s, it has enjoyed a revival since the late 1980s, due directly to the work of master brick mason Gerard Lynch, and is now flourishing. Although manifestly fulfilling a technical purpose, it is discussed in this section under decorative features as the use of gauged brickwork, rather than employing standard bricklaying practices, was an aesthetic choice.

The bricks used to create gauged brickwork are different in consistency and texture from those used in the majority of traditional brickwork. Only the topmost, geologically younger brickearth and clays, with a naturally high silica level, are used. Until the middle of the 19th century, the chosen material was from downwash alluvial deposits so some inclusions can be seen. After this time, with improvements in brickmaking technology, those materials would be washed and screened to be free from inclusions such as the grit or pebbles that can be found in some handmade bricks as seen in Figure 6.8. By the second quarter of the 19th century, the bricks used for gauged brickwork were often made slightly larger, or "oversized", than would commonly be the case. This development was to allow the bricks to be cut accurately to shape using a bow-saw with a twisted wire blade within a suitably profiled "cutting box", and carefully finished by rubbing to the precise shape and size while remaining in proportion with the surrounding brickwork. These rubbing bricks were, and still are, "low-fired" below 900°C within the kilns, which means they do not have the "fire skin" found in other bricks; it is this characteristic that allows them to be cut and rubbed to shape and profile, presenting the same texture and appearance throughout, without detriment. It is a fact that the traditionally made rubbing bricks used for gauged brickwork are generally more porous and softer, but true rubbers have interconnected pore structures, which means they cope very well with the action of frost. That said, gauged brickwork is potentially more vulnerable to damage than most brickwork, particularly on exposed features like decorative cornices and other projecting enrichments that lack properly detailed leadwork, and from inappropriate abrasive masonry cleaning systems.

Gauged brickwork was traditionally set using what is termed a "fine-stuff" mortar and based on either the traditionally preferred feebly hydraulic, "Grey" or "Stone" lime putty or, less commonly, a non-hydraulic or "air" lime putty. Preparing the lime binder like this allowed for the very fine joints of finely screened lime putty mixed with a suitable proportion of aggregate of a very small size of grain, often referred to as "silver sand". When surveying a brick building, gauged brickwork can be identified through its characteristic use of thin mortar joints and very accurately cut and rubbed bricks. Where gauged brickwork is identified during any survey work, this should be carefully noted, and the feature recorded. Gauged work adds considerably to the heritage significance of a building and will also have particular requirements if the building is being repaired and maintained.

Gauged brickwork can be found in a number of different contexts on buildings, such as aprons, platt bands, cornices, niches and pilasters, but is most commonly seen employed to various forms of arches. Gauged brick arches can be seen on buildings where the main body of the surrounding brickwork is standard facing brickwork. It should be noted that some buildings, particularly in the late Stuart and early Georgian period, used "ashlared" gauged brickwork for their entire façades, a fashion that was revived on some high-status brick buildings during the late 19th century. Considerations of its great cost, however, saw the use of gauged brickwork restricted increasingly to specific areas of buildings in these later periods.

The defects that can affect gauged brickwork are broadly those that will be seen in other types of brickwork, as discussed in Chapter 6. Open joints between individual bricks should be noted. Extra care is likely to be required to fully determine whether mortar joints are open, given their slenderness. Defects with rubbing bricks themselves may be as a direct result of previous inappropriate interventions. The soft nature of rubbing bricks makes them particularly

Fig. 4.3 Gauged brickwork showing fine joints and cut and rubbed bricks to impressive effect (copyright Georgian Brickwork).

vulnerable to damage by ill-advised aggressive cleaning methods, for example. Like all traditional soft bricks, rubbers are also vulnerable to defects that arise as a result of the use of inappropriate modern mortars, and particularly those that use Ordinary Portland Cement (OPC) as the binder.

All remedial works on features of gauged brickwork, whether conservation, repair or restoration, must only ever be undertaken by first-class brick masons with proven knowledge, skill and years of meaningful experience on this high-end branch of the craft; and those who will utilise materials, tools and methods wholly empathetic to the period of construction. Gauged brickwork is a uniquely special and hugely important part of the heritage of traditionally constructed brickwork, and therefore requires particularly informed level of special care in its correct survey, understanding and assessment. This short section is just an introduction to what is undoubtedly a quite complex subject, and a number of detailed works on the assessment and repair of gauged brickwork are noted in the References and Further Reading section at the end of this chapter, with particular attention being drawn to Lynch (2006a, 2006b, 2007).

4.3 Polychromatic brickwork

Polychromatic brickwork employs bricks of different colours for decorative effect. The term is most commonly used to refer to brickwork dating from the mid- to late 19th century, but it can apply to any brickwork that purposefully uses bricks of differing colours for decorative purposes. The technique came into common use in the United Kingdom following advances in brick manufacturing, with one technical author on brickmaking noting in 1850 that "ornamental bricks might be made of a variety of colours" and that there was a rising taste for polychromatic decoration (Dobson 1850), with several other works around the same period recommending the use of contrasting colours in brickwork.

Polychromatic brickwork generally uses lighter colours than are used on the main face of brickwork to highlight building features. White, cream or yellow bricks are used particularly around window reveals, to form arches and at the quoins of buildings. Lighter bricks are also commonly found forming cornices and string courses. Simple patterns were generally used, but in some cases more complex setting out of polychromatic work occurred. This can be seen in Figure 4.4, where buff bricks are used to pick out what is termed a "Greek key pattern" in the brickwork. Polychromatic brickwork is used extensively in the building featured in Case study 4.

In some cases, darker coloured bricks were used to decorative effect. Black and blue predominate the use of darker bricks, and these were sometimes used in conjunction with the lighter coloured bricks described above. Where polychromatic brickwork is employed in a decorative context, this can see the bond pattern of the brickwork disrupted, as in Figure 3.50. If such disruption to bond is seen, this should be recorded during the survey.

Fig. 4.4 This building presents a complex scheme of polychromatic brickwork. A survey would note that polychromatic decoration utilising a Greek key pattern is used at cornice level. This is formed in white brickwork, with the main body brickwork being dark red in colour. At least two colours of bricks will therefore be required for any repair work to this area of the building.

Fig. 4.5 Despite common associations with large industrial buildings, polychromatic brickwork can be found in a wide range of structures, including relatively humble domestic structures such as this.

Fig. 4.6 An unusual example of the use of darker as well as lighter bricks in polychromatic brickwork; the setting out of the bond to accommodate the decorative scheme is very complex as the building is a former industrial chimney, Cox's Stack, Dundee.

The principal defect likely to be encountered with polychromatic brick-work is a failure of the bricks themselves, as discussed in section 6.3. There are no specific failures that are likely to be seen in polychromatic brickwork, and when surveying polychromatic work the common defects noted in chapter 6 may be encountered. There is generally no specific vulnerability in the lighter or darker bricks used to create polychromatic effect, and in many cases these bricks are more durable than those used in the main body of the brickwork. When assessing a structure that incorporates polychromatic brickwork, recognising this and accurately recording how it is used is likely to be the most significant action required. The number of polychromatic brick buildings is in decline as industrial buildings are demolished, so the use of such decorative work is of particular heritage significance and should be recorded where found.

4.4 Special and glazed bricks

The use of bricks other than the standard rectangular prism is a further way of adding decorative embellishment to brickwork. Bricks, either cut to shape or moulded to a particular requirement, have been discussed in section 3.2.3, where they are put to a particular technical purpose. It should be noted, however, that as well as the practical and technical uses to which special-shaped bricks were put, the decorative uses of special bricks were considerable. Examples of the use of special bricks in a decorative context can be seen in Figure 4.8, and also in the case study at the end of this chapter. Where special bricks or bricks cut to shape have been used decoratively, this should be recorded carefully during assessment.

The earliest form of shaped bricks used in a decorative context saw bricks cut or carved to shape post firing. This would have been achieved using the brick axe, as discussed in section 3.2.3. Advances in manufacturing techniques, especially the ability to press bricks to shape in the 19th century, allowed special shaped bricks to be moulded to shape prior to firing. This provided a much greater diversity in the shapes of bricks that could be supplied, with specials often incorporated into the polychromatic decorative schemes discussed above. Some decorative special shaped bricks were used in the formation of arches, as seen in Figure 4.22. Special bricks could also form cornices in a decorative fashion. While these bricks still fulfilled their technical purpose, either as parts of a wall or as arch voussoirs, the decorative aspect of their use should not be overlooked. Special shaped bricks are an integral part of the aesthetic of many traditionally constructed brick buildings. Examples of the use of special bricks in a decorative context include double splayed bricks used to form dogtooth courses, bottle nose bricks used to form decorative quoins, arches and cornices and small beaded bricks used to form decorative arch voussoirs. Where specials are used, these should be recorded, where possible naming the shape used, although it should be noted that some were produced as bespoke products for a particular building, generally referred to as "special mouldings".

Glazed bricks are also used in a decorative context in traditional brick-work. There are two principal ways in which glaze has traditionally been applied to bricks: salt glazing and enamelling. "Common" salt glazed bricks were pressed bricks which had salt thrown into a kiln at high temperatures, which then vaporised and reacted with the surface of the brick. Where what was termed "best" salt glazed bricks were formed, these were dipped into a slip of fine clay before being salted and fired. Enamelled bricks were first fired and then dipped into a glaze before being fired a second time. Sometimes what was known as the wet dip process was used; this omitted the prelim-inary firing.

The colour of such glazed bricks could be altered by admixture of ingredients to the glaze. One technical author in the 19th century noted that, "Ornamental bricks are now made in many designs and of a great variety of colours, which are usually produced by the employment of metallic oxides and sometimes ochreous metallic earths, oxides of lead, platinum, chromium and uranium are used for very fine colours" (Davis 1889, p. 71). Glazed bricks are generally found in later 19th century brickwork, used for both technical and aesthetic purposes.

When used decoratively, glazed bricks are often employed to pick out patterns within brickwork constructed of other brick types. Where glazed

Fig. 4.7 Green and orange glazed bricks used decoratively in a façade; a decorative bond is also seen.

bricks are found in a decorative context, this should be noted during survey and assessment. It should include a description of how and where the glazed bricks were employed as, should the survey be used to inform future repair work, the glazed brickwork can be vulnerable to inappropriate interventions such as cleaning.

Where glazed bricks are used over a wider area (more commonly a functional rather than decorative choice), generally only the front face of brickwork

Fig. 4.8 Glazed brick has been used to considerable decorative effect in this building despite its utilitarian purpose as a public toilet.

Fig. 4.9 In this case, the glazed bricks have become dislodged, revealing a backing of common brick where structural problems are encountered with a wall faced with glazed bricks; differential performance or insufficient bonding with the backing brickwork may be an explanation.

is formed of glazed bricks, with the backing material being a different, cheaper form of brickwork as seen in Figure 4.9. Glazed bricks were generally produced with either the stretcher or the header face glazed – or, in some cases, two glazed faces. Where glazed brickwork is backed by a different brick type, the two materials delaminating from each other may be a source of weakness and cause of deterioration.

Glazed bricks themselves can suffer from the various decay mechanisms discussed in Chapter 6. There are, however, some specific considerations around glazed bricks and their assessment. While glaze can fail and spall from the surface of bricks, glaze can also suffer what is termed "crazing". Where found, this should be recorded, but does not necessarily mean that the brick has failed as it can occur during or soon after manufacturing. The most significant defect likely to be encountered where glazed brickwork is used decoratively is a lack of bonding or movement between the glazed façade and the backing brickwork, as seen in Figure 4.9.

4.5 Dogtooth and dentil courses

Dogtooth and dentil courses are decorative embellishments incorporated into traditional brickwork. When applied to brickwork, the term "dogtooth" refers to the use of bricks turned through 45 degrees so as to present a corner on the face of the wall. A dogtooth course can be formed of a standard brick turned to the appropriate angle, but it can also be formed of special bricks. It is common to see dogtooth courses used in conjunction with polychromatic brick decoration. The most common application for dogtooth brickwork is to form string courses or cornices, but it can also be incorporated into arches, as seen in the case study at the end of this chapter.

A dentil course is formed in brickwork by allowing some header bricks to project from the face of a wall to form a tooth-like projection. This is well summarised by Pasley (1826, p. 222), who notes it as being formed "by causing part of the headers of one course of bricks, the last but one, to project a quarter, or sometimes half a brick from the face of the wall". Dentil courses are generally used at cornice level and can be formed of special bricks as seen in Figure 4.11.

Neither dogtooth nor dentil courses present much in the way of specific survey requirements. Where they are used, this should be noted as being part of the overall aesthetic of the brickwork being surveyed. In terms of defects, other than the usual defects exhibited by brickwork, the only particular consideration with dogtooth or dentil courses is that these can create small ledges on which water may lie. This may lead to decay if significant water ingress is taking place. As this is likely to be a problem only if there are wider failures within a building, dogtooth and dentil courses should generally only present a requirement to be noted as part of an overall decorative scheme.

Fig. 4.10 Dentil course in a 19th century industrial building.

Fig. 4.11 Detail of dentil course formed of moulded convex special bricks.

Fig. 4.12 An unusual example of a double dogtooth course.

4.6 Historic paints and surface finishes

While brickwork presents a durable surface finish as built, it is also sometimes treated with an applied surface coating. This could be for either decorative or technical reasons, and includes paints, renders and modern surface coatings such as water repellents, as considered in Case study 6.

It should be noted that some paints and surface finishes were applied during or soon after original construction and therefore are of heritage significance. The use of paint or colour wash on brick façades can be seen in the United Kingdom from the late 15th century. These early coatings were primarily used to conceal variations in the colour of bricks and also in mortar joints, which could be asymmetrical. It is rare for these early paint finishes to have survived, and where they are encountered during survey work, it may be appropriate to highlight these to the relevant statutory heritage authority. Following the application of colour wash, it was also common practice to line out some joints in a technique known as "pencilling". The use of paints and colour washes is an important aesthetic component of brickwork. Historic paint finishes used a variety of natural pigmentations such as copperas and raddle. Limewash may also be found utilised as a coating to traditionally constructed brickwork.

Some brickwork was designed and constructed in a way that was intended to be covered with an external render. This could be for both technical and aesthetic reasons. There was, in the Georgian period, a fashion to render brickwork, which was sometimes lined out to imitate stone in some parts of the United Kingdom. In the latter part of the 19th century, bricks manufactured of colliery shale were produced with the intention that they would be rendered. This can mean, if the render is removed, the bricks may not be of suitable durability to withstand water ingress or the actions of frost Where brickwork was intended to be rendered and the render is failing, this should be noted in survey work as the bricks underneath may be suffering from decay. The removal of render to allow for inspection of what is underneath is an invasive process, however, and may require statutory consent or adherence to archaeological considerations.

Where bricks that were originally intended to be rendered have come to be exposed to the elements, they may exhibit exacerbated or accelerated decay. Signs that brickwork was originally intended to be rendered include the presence of small patches of render surviving, such as under the eaves at roof level, and also deeper margins around windows, which would have been created in order to allow the render to finish flush with the margin. Where it is suspected that brickwork was originally rendered, this should be noted in a survey as the bricks may be more vulnerable to decay.

Conversely, renders, paint and surface coatings using modern materials can have a significantly deleterious effect on traditional brickwork. If Ordinary Portland Cement is used as the binder, external renders can have the effect of trapping moisture within brickwork, in a similar way to the effect of cement mortar discussed in section 6.10. Moisture is always likely to penetrate behind cement-based renders, either through large cracks or, more commonly, through micro cracks that draw moisture into the brickwork by capillary action. When this moisture has entered the brickwork behind the render, it is often hard for the moisture to dissipate, saturating the brickwork and creating conditions that severely impact bricks, render and associated elements, such as timber joist ends. Where cement render has been applied to brickwork, particular care should be taken to note any sections that have become delaminated from the brickwork or that are bossed, as there is a high likelihood that the bricks behind may be suffering from deterioration.

Other external coatings, such as masonry paint, may also trap moisture in brickwork, with this moisture leading to the blistering of paint. A recent potential cause of deterioration in traditional brickwork is the use of impermeable insulation externally on brickwork. This can have a similar effect to the use of cement-based renders, trapping moisture within the brickwork, leading to decay of the brickwork, the mortar and any timber that may be built into the brick walls internally. Any external coating, be it paint, render or insulation, should be noted in survey work, and an attempt made where possible to ascertain any possible damage to the bricks behind it.

Bricks may have been coated in a water repellent or other surface coating designed to improve their durability. These modern products are not a substitute for a well-maintained building and can have the effect of trapping moisture within brickwork. Recognising these surface applied materials can be difficult during survey work, as can be seen in Case study 7. Such surface applied materials may give an impression of surface decay of brickwork or of being

Fig. 4.13 Modern cement renders can cause significant damage to traditional bricks. In this case, the relatively strong facing bricks have suffered only minor decay although the efflorescence in the right side is likely to be caused by the render.

Fig. 4.14 This image clearly shows the extent to which modern cement render can obscure the aesthetic of traditionally constructed brickwork. The image on the right shows the building following the removal of render (copyright Dublin Civic Trust/Ros Kavanagh).

some form of surface pollutant. It will be seen from Figure 6.37 that where water repellents or other coatings of this nature are found on brickwork they will often delaminate as a thin layer of material. Identifying such finishes can be difficult, and it may be that recourse to records of previous interventions to a building is required. It may also be possible in some cases to have such surface finishes analysed in an attempt to identify them using techniques set out in Chapter 6. Where water repellent has been applied to brickwork, it is likely that the bricks suffered from decay at some point, explaining the decision to apply the water repellent in the first place. The water repellent is unlikely to have addressed underlying defects, which is something that may need to be assessed further.

4.7 Terracotta and faience

Terracotta and faience are both architectural ceramics commonly found used in conjunction with traditional brickwork. Both materials are formed from clay ground and mixed with sand or powdered fired clay, which has enough plasticity for it to be moulded and shaped. Such materials are generally harder and fired to a higher temperature than a standard clay brick would be. The difference between terracotta and faience is primarily that faience is covered with one or more glazes, whereas terracotta is not. The glazing imparted on faience is sometimes used to create the effect of stone. Although some firms produced only terracotta and faience, many companies produced both architectural ceramics and bricks.

Terracotta and faience were a cheaper alternative to carved stone, as many units could be cast from the same mould. Terracotta and faience can be found as a cladding or facing material on buildings. Where this was the case, both terracotta and faience were often attached to brickwork by metal ties or anchors, with the ceramic material being "hung" on the load-bearing structure. Units were often manufactured hollow to make them lighter and easier to use. Sometimes these hollow units were manufactured with what was known as webbing, a series of reinforcements to make the unit more rigid.

Terracotta, faience and bricks have a close relationship in traditional construction. Terracotta and faience are commonly found used as embellishments to brickwork, with brickwork also being used as the principal structural component in terracotta structures. Differentiating between terracotta and faience and stone embellishments can be difficult unless close inspection is possible. Where small areas of surface loss occur, as in Figure 4.17, this will reveal whether a fired clay material is underneath.

A wide range of defects can affect terracotta, faience and brickwork, but this is beyond the scope of this book; the reference section of this chapter provides abundant sources of further information. The principal defects that affect architectural ceramic may be broken down to a deterioration of the material itself and a failure of the anchoring system. Structural deterioration of the building to which the material is attached and bonded can also cause damage to faience and

Fig. 4.15 Terracotta elements such as this are common embellishments on traditional brick buildings; differentiating them from special bricks generally comes down to size, although terracotta was often manufactured hollow.

Fig. 4.16 Faience units such as this were often manufactured in such a way as to give the appearance of carved stone but at a lower cost.

terracotta. This is particularly true where the material is attached to a ferrous metal structure that is suffering corrosion as this will cause "oxide jacking" – the process of expansion of metal when corroding. As the anchoring systems that attach terracotta and faience architecture to buildings are hidden from view,

Fig. 4.17 Failure of terracotta is similar in many ways to the decay process for bricks; here small areas of spalling have revealed the fired clay beneath.

Fig. 4.18 In this building, terracotta has been employed to create various parts of the façade, with high-quality facing brick being used as well. The use of terracotta in this way is a common feature of late 19th century brickwork (copyright Jill Fairweather).

such deterioration is often hard to detect unless a serious problem has occurred. Any evidence of displacement of elements, spalling or rust staining should be followed up with a thorough investigation to check for potential failure of anchoring systems, as this can potentially be a very dangerous situation.

Where terracotta and faience are found used in a traditionally constructed building, this should be recorded in the survey and consideration given to any defects that may be present. Architectural ceramic can be part of the heritage significance of a building and can also present particular defects. These may require further specialist investigation; the repair and maintenance of terracotta and faience deserve a book of their own.

4.8 Case study 4: Kinning Park Colour Works, Glasgow, United Kingdom – assessment and recording of decorative features

The building that forms the basis of this case study is the Kinning Park Colour Works in Glasgow. Built in 1895, the small building formed the offices of a firm of paint manufacturers. Although small in scale in comparison with many late 19th century brick industrial buildings, it presents a wide variety of decorative features, making it an ideal case study when considering the survey and recording of decorative brickwork.

The most immediately obvious decorative feature is the use of polychromatic brickwork. This sees white brick contrasting with the orange-red facing bricks employed for the majority of the façade. Polychromatic brickwork is employed at several points, forming string courses, oculi, quoins and arches above both windows and doors, and at window reveals. Such extensive use of polychromatic brickwork is unusual and worthy of note.

When recording polychromatic brickwork, it is important to note not just where it is used but also the manner in which the polychromatic brick may affect the rest of the masonry. For example, polychromatic brickwork can in some instances disrupt bond patterns. The way the polychromatic brick is constructed should also be recorded. A complex example of this can be seen in the intricate archway above the main entrance to this building, this being formed of alternating red and white brick.

Although arches are examined in detail within the examination of technical characteristics in Chapter 3, it should be noted that arches often form part of decorative schemes, so are also relevant here. This case study provides an excellent example of this. The semi-circular arch above the main entranceway is clearly both decorative and functional. To consider the technical aspects of the arch, it is five and a half bricks in depth but of complex and decorative configuration. The arch is formed of concentric courses of header/stretcher/header/stretcher/header/header/header/stretcher. This arrangement incorporates a heading course of special bricks and two heading courses of dogtooth brickwork, which can be considered as purely decorative. The achievement in setting out such a complex arch, particularly given the other elements of the façade, is also worthy of note. The majority of this is decorative rather than functional, and the structural requirements of the arch above this entrance could have been met through the use of simple concentric rings of header bricks. It can thus be seen that arches such as this can form both decorative and functional aspects of traditional brick buildings.

The construction of several oculi windows is a further example of both functional and decorative brickwork coexisting. While serving a functional purpose, the oculi in this case are certainly evidence of decorative treatment, and the use of oculi rather than more regular-shaped rectangular windows is a choice made on the basis of aesthetics rather than functionality. The oculus in the front façade at Kinning Park is one and a half bricks in depth, formed of an outer course of headers and an inner course of special stretchers. The special bricks in the oculi are of the small-beaded form.

At three points in the façade, dogtooth courses are formed. This can be seen forming a string course at first-floor level with two dogtooth courses incorporated into the arch above the principal entrance. A course of dogtooth brickwork is also used at cornice level; this is likely to be formed of special bricks.

The use of special bricks in this building is largely decorative in function. Moulded convex bricks are used at several points in this impressive construction, with a course of single splayed bricks also used in the arch. A combination of single splayed and moulded convex bricks are used to impressive effect at the doorway, as shown in Figure 4.21. Single splayed bricks are also used to form

Fig. 4.19 The impressive decorative treatment applied to this 19th century industrial building is complex to assess and record.

Fig. 4.20 Single splayed bricks used to form the window sill with a moulded convex also used at the reveal.

Fig. 4.21 An intricate arrangement of single splayed bricks is used to form the reveal for the main entrance to the building.

Fig. 4.22 Detail of the exceptionally decorative arch at Kinning Park Colour Works showing the use of specials, polychromatic brickwork, dogtooth courses and special bricks.

the window sill in this building. Small-beaded bricks are used to form the oculi, with moulded convex used to form the projecting semi-circular arch above. The dogtooth courses are formed of double splayed bricks. This represents an impressive range of uses of special bricks in a decorative context and shows how a detailed understanding of the range of specials used traditionally can inform survey and recording.

Lastly, it should be noted that sandstone is also used in a decorative context in this building, forming pillars and an arch above a secondary doorway. Although beyond the scope of this book, in a comprehensive survey of the building the use and condition of these stone elements should be undertaken.

The Kinning Park Colour Works provides an excellent example of how traditional brickwork can incorporate decorative elements. These include the use of special bricks, dogtooth courses, polychromatic brickwork and decorative construction for arches. The correct recording of these decorative elements is critical to fully appreciating the heritage significance and overall construction of this building. It serves as an illustrative case study of the importance of recording both decorative and functional brickwork in traditional buildings.

References and further reading

Ashurst N. (2008), *The Investigation, Repair and Conservation of the Doulton Fountain, Glasgow,* Edinburgh: Historic Scotland.

Darbishire H. (1865), "The Introduction of Coloured Bricks in Elevations", *The Civil Engineer and Architects Journal,* vol. 28, pp. 66–9.

Davis C.T. (1889), *A Practical Treatise on the Manufacture of Bricks, Tiles, Terra-Cotta etc.,* Philadelphia: Henry Carey Baird.

Dobson E. (1850), *A Rudimentary Treatise on the Manufacture of Bricks and Tiles,* London: John Weale.

English Heritage (1994), "The History, Technology and Conservation of Architectural Ceramics", *Conference Papers, UKIC/English Heritage Symposium,* London: English Heritage.

Lynch G. (1994), *Brickwork: History, Technology and Practice,* London: Donhead.

Lynch G. (2006a), "The colour washing and pencilling of historic English Brickwork" *Journal of Architectural Conservation,* vol. 12, no. 2, pp. 63–80.

Lynch G. (2006b), *Gauged Brickwork: A Technical Handbook',* London: Routledge

Lynch G. (2007), *The History of Gauged Brickwork,* London: Elsevier

Pasley C.W. (1826) *Practical Architecture,* Chatham: Royal Engineering Establishment.

5 Assessing age and significance

Ascertaining the age of brickwork during survey and assessment fulfils a number of purposes, depending on the nature of the survey being carried out. It can aid in ascertaining the heritage significance of a building or a given part of a structure. Where, for example, brickwork from an earlier period survives behind later alterations, it is possible that the structure is of greater heritage significance than may first have been supposed. In some cases, it may be found that a section of brickwork represents a later alteration to a structure and can therefore be used to identify distinct phases of development.

5.1 Identify age: Heritage significance and defects

The age of brickwork can prove informative when identifying defects and their cause. To take one example, where horizontal cracks are found in brickwork, knowledge of how construction has developed can be beneficial. If such horizontal cracking is found in brickwork dating to the mid-18th century, decay of bond timbers may be a possible explanation. If the brickwork dates to the latter part of the 19th century, such horizontal cracks may be a result of decaying cavity wall ties. If a brick structure was constructed in the 17th century, decay of hoop iron bond, which did not come into common use until the 19th century, is unlikely to be a cause of the defects. Whilst further investigation of the cause of these cracks would be required to give a definitive answer, knowledge of the date of construction and the methods and materials used in a given period can provide clues to aid survey work. Likewise, if brickwork was constructed in the 18th century, it is more likely that a front façade of facing bricks being applied to a backing of common brick was used. A lack of proper bonding between the two can lead to problems of delamination and ultimately structural instability. A correct understanding of the period in which brickwork was constructed is therefore critical for the correct identification and interpretation of defects.

It is only possible to deploy such knowledge when interpreting brickwork with an understanding of how bricks and brickwork developed over time. This is vital to interpreting brick structures and their age. This chapter seeks to provide guidance on the development of brickwork during certain specific time

DOI: 10.1201/9781003094166-6

periods and to highlight characteristics that can aid in the dating of brickwork. Changes in brick type, size, gauge and bond can all aid in ascertaining the age of brickwork. An understanding of developments within brick construction is a valuable aid in ascertaining the period in which a piece of brickwork was constructed, but this must be viewed in the overall context in which the brickwork sits. Understanding the wider context of the building as a whole will also aid in ascertaining the age of brickwork. If the broad period of construction is known, this provides a baseline for dating brickwork. However, as is always the case with traditional brick buildings, expecting the unexpected is a good maxim from which to work.

When surveying historic brick buildings, it will often be found that several different phases of development are evident. Over the lifespan of any building of traditional construction – which by definition will have been extant for over 100 years – additions, alterations and demolition will change the character and appearance of the structure. Survey work will often involve the identification of different phases of development. This can be a complex process, and the services of a specialist building pathologist – or, for certain buildings, an archaeologist – may be required. In some cases, it will be obvious how a building has evolved over time; in others, this will be more complex to unravel.

The focus of this chapter is the development of brickwork within the United Kingdom. The most important factor in successfully ascertaining the age of brickwork, however, is common to all parts of the world where brickwork is found: a strong knowledge and understanding of the brickwork of the country, city or territory where it is constructed. Local knowledge and context are vitally important. Brickmaking and the use of brick were often influenced by localised factors. For example, the use of brick in Scotland followed a very different course from that in neighbouring England, with brick not becoming a widespread material in Scotland until the later part of the 18th century. This was largely due to the presence of easily obtainable stone and a strong earth building tradition at the vernacular level. Knowledge of this makes the dating of brickwork in Scotland easier. Within this, however, it is known that brickworks were operational in some parts of Scotland in the early 18th century, so it is not a surprise to find early brickwork in Perthshire during survey; however, to find it in the Highlands would be. In China during the Ming Dynasty period, brick kilns were established and brickmakers were obliged to inscribe each brick produced as a method of monitoring brickmaking quality. Provenance of bricks can therefore be traced by studying these inscriptions. Knowledge of the development of brickwork in a specific country or region can therefore be seen to be of considerable assistance when ascertaining the date of a particular piece of brickwork.

Identifying the age of brickwork is often of greatest importance when trying to ascertain its heritage significance. The significance of a building may be recognised by a local or national heritage agency. Historic Environment Scotland, English Heritage and other national bodies, both in the United Kingdom and around the world, maintain lists that record buildings of particular heritage significance. On a global scale, organisations such as UNESCO also recognise important heritage structures, a number of which are constructed of brick. In most cases, however, brick buildings will not be designated as being of significance by either national or international bodies. This is where it becomes necessary for building professionals to assess significance when surveying traditional brick structures.

Fig. 5.1 Not every building will be as clear in showing where an alteration has been made. The brickwork, which previously was internal, is of very poor quality and is likely to be vulnerable to decay and water ingress.

Fig. 5.2 The heritage significance of a building can include consideration of craft practices; in this case, the pointing technique is a unique surviving example of a Scottish version of tuck pointing.

There are many different aspects of a building that can contribute to its significance. This may be due to methods used in construction or examples of unusual craftsmanship. Associations with historical events or people may also add to a structure's significance. This is something that can be difficult to appreciate during survey work and can throw up some interesting examples, such as that in Case study 5. A number of guides and resources are available to help ascertain the significance of a traditional building. When surveying a building of traditional brickwork, identifying the technical and aesthetic features described in Chapters 2 and 3, as well as the age of the structure, will aid in ascertaining the heritage significance of a building. There are also various standards that exist to help assess the development of a building and its heritage significance. The recent British Standard regarding heritage significance and impact is a helpful guide to surveyors in this complex area.

The following sections provide a broad chronology of brickwork in the United Kingdom with indicative examples of technical and aesthetic aspects that can aid in the dating of brickwork. For each period discussed, however, a wealth of further information is available and should be referred to in order to broaden knowledge of the development of brickwork to aid survey work.

5.2 Roman, Anglo–Saxon and Medieval brickwork

The skills required to manufacture fired clay bricks first travelled to Britain when large parts of the country became part of the Roman Empire in the 1st century CE. The bricks made by the Romans were generally wider and thinner than the bricks we see today. These bricks were used in various ways, including as lacing courses in walls of rubble stone masonry and in the construction of supporting pillars for hypocaust heating systems. Roman bricks can be found throughout the United Kingdom at significant sites such as Burgh Castle, Norfolk and as far north as Newstead in Scotland.

Today, Roman bricks are most likely to be found in an archaeological context; however, they can be found reused in later buildings. When the Romans left Britain in the 5th century CE, it is thought that brickmaking came to an end until the 12th century. Some buildings from this brickmaking interregnum period saw the reuse of Roman bricks as at the nave of St Alban's Abbey where construction began in the 11th century reusing materials from Roman Verulamium.

The earliest known use of brick manufactured in the United Kingdom after the Romans left is widely regarded as being at Coggeshall Abbey in Essex, dating to 1190 CE. Beverley North Bar in East Yorkshire is a very good surviving example of Medieval English brickwork, with construction beginning around 1409. Here the bricks were thin (50 mm) and, in common with indigenous bricks of the time, somewhat uneven in shape and size. This gave brickwork of the time a distinctive character with wider mortar joints and uneven bonding. Some fine and significant examples of brick buildings from the medieval period survive, including Rye House Gatehouse (built around 1443), and Thornton Abbey, Lincolnshire (built around 1382).

Fig. 5.3 Roman bricks used as a lacing course in Verulamium. Roman bricks from the city were reused in the construction of St Albans Abbey (copyright David Pickles/Historic England).

Fig. 5.4 Late medieval brickwork, England. Technical features of note include irregular bond and wide mortar joints to account for variations in the size and shape of the bricks (copyright SPAB).

A decorative feature evident in medieval brickwork and extending into later brickwork is the use of diaper patterned brickwork as discussed in section 4.1. This saw patterns of saltires and diagonal intersecting lines picked out using well-vitrified or "flared" headers, which were over-burnt during firing and presented a grey to blue colour. Both cut and moulded bricks were used in this period to add decorative embellishment. Medieval brickwork was built with considerable skill, and was used to create vaulting, circular and angled walls as well as arches, thereby demonstrating a good understanding of the properties of the material. It should also be noted that in this period brick was a material with relatively high status attached to its use, which saw its employment largely confined to those at the upper echelons of society in both church and state.

5.3 Tudor, Elizabethan and Stuart brickwork

Following on from the early brickwork of the Roman and Medieval period, a greater use of the material can be seen in the 16th century. This includes such majestic buildings as Layer Marney Tower (built in 1523) and Hampton Court Palace. Brickwork at this time began to exhibit greater standardisation in terms of technicalities, including greater use of recognised bonds – most commonly English or English Cross bond, although irregular bonding was still being used. The high status of many of the buildings being built in brick is reflected in the use of decorative features such as diaper work and in the use of techniques such as colour washing and pencilling of joints, as discussed in section 4.6. Bricks of this period continued to be somewhat irregular in size and shape. Some of the brickwork at this time was of an exceptional standard, however – for example, at Leez Priory, Essex (built in 1563) and the moulded brickwork at Roos Hall, Suffolk (built in 1583).

The 17th century saw a further expansion in the use of brick. Buildings such as Balls Park, Hertford (built in 1640), are indicative of brickwork at this time. With regard to technical features, the use of Flemish bond can be seen to gain popularity. A highly significant development in this period is the emergence of gauged brickwork as an enrichment discussed more fully in section 4.2. An early example of such gauged brick enrichment is the Old Meeting House, Norwich (built in 1693). The Great Fire of London in 1666 saw considerable use of brick in the rebuilding of the city. The great gathering of craftsmen occasioned by this rebuilding was to have a long-term impact on brickwork in the United Kingdom. Previously a material used mainly for houses of the powerful and wealthy, by the end of the 17th century brick was becoming much more commonly used for buildings of many different classes. The geographical spread of brickwork also expanded at this time, with brick buildings becoming more common throughout England and Wales.

Fig. 5.5 A and B Hampton Court Palace is one of the most significant brick buildings of the Tudor period, with brickwork of exceptional quality (copyright David Pickles / Historic England).

Fig. 5.5b

Fig. 5.6 Diaper patterned brickwork, Ipswich; the date stone reads 1549 but not all brickwork is so easily dated (copyright Mathew Slocomb).

5.4 Georgian brickwork

The Georgian period saw considerable expansion in the use of fired clay bricks in the United Kingdom. Brick was used to create many architecturally impressive buildings such as Chicheley Hall, Buckinghamshire (built in 1719) and Bailey Hall, Hertford. The Georgian period also saw brick used for terraced housing in a way that had previously not been seen, as at Union Place, Wisbech (built around 1800). These buildings often incorporated gauged brick enrichment as the quality of bricks and the skills of craftsmen both improved.

Fig. 5.7 An example of Georgian brickwork, Cadogan Place (copyright Georgian Brickwork).

It should be noted that at this time brick was also used to build housing for both rural and urban workers as well as the houses of the privileged. This expansion saw brick used across a greater geographical area, with considerable expansion in Scotland in the 18th century for example. The Georgian period also saw brick used to construct industrial buildings throughout Britain, with early textile mills often being constructed partly or wholly of brick. Stanley Mills in Perthshire is a good example. This included the use of brick arched flooring to build fireproof mills as at Ditherington Shropshire (built in 1797) before quickly spreading throughout Britain. Rural housing across all social classes also began to be constructed of brick, in some cases reflecting influences from vernacular construction techniques, as at Flatfield Farmhouse, Perthshire (built in 1785), discussed more fully in Case study 5. Advances in brick manufacturing in this period included the more widespread establishment of permanent brickworks as opposed to temporary brickfields. Despite this, brickmaking remained an operation that did not employ mechanisation apart from the limited use of horse powered pugmills. Societal change in the form of urbanisation and industrialisation also influenced the use of brick, with an increase in the applications to which the material was put. Although not on the scale seen in the later railway construction boom, the construction of canals and infrastructure to support trade, such as warehouses, also saw an increased use of brick.

Fig. 5.8 Gauged brickwork such as this came into more common use in the Georgian period; this example was reconstructed during repair work using traditional craft practices (copyright Georgian Brickwork).

Fig. 5.9 The use of brick to form a groined vault in an 18th century military structure; brick was used in this context despite other parts of the building being
constructed of stone masonry.

Technical features of this period include the common use of Flemish bond,
although irregular bond was common in rural buildings. It was at this time
that gauged brickwork became more widely utilised in British brick buildings.
Brick was often used in the 18th century for specific purposes within structures
of stone. Such uses include vaulting, lining ashlar walls and the construction of
internal partitions.

Overall, the Georgian period was one of considerable expansion in the use
of brick. Production rose considerably and the types of buildings constructed
of the material also widened. Brick was used to build structures from the most
prestigious to the humblest, and every type in between. Exhibiting a range of
technical and aesthetic features, Georgian brickwork in the United Kingdom is
both diverse and durable.

5.5 Victorian and Edwardian brickwork

The early part of the 19th century saw the use of brick continue in much
the same way it had in the 18th century. This would change, however, as the
1800s progressed. The 19th century could rightly be considered a century

built of brick. With the railway mania that overtook the country in the mid-19th century, the need for bricks to construct bridges, line tunnels and ancillary buildings saw a considerable expansion in the use of the material. Other engineered structures constructed of brick at this time included tall chimneys, lighthouses and an incredible range of industrial structures. This use of brick in an engineered context has left a considerable legacy in the United Kingdom from North Unst Lighthouse, Shetland in the North to St Pancras Station, London in the South.

Brick was used throughout Britain at this time to construct vast rows of housing for industrial workers. From red brick terraces in the North of England to tenements of yellow London stock bricks in the South-East; and from rows of miners' cottages of colliery shale brick in Scotland to the dark red brickwork of Belfast, the workers in the industrialised Victorian United Kingdom were housed in structures of brick. The material was also used to construct the mills, factories and tall chimneys that supported the industrial enterprises driving the country forward.

Considerable change in the way brickmaking was undertaken was experienced from the mid-19th century onwards, as set out in Chapter 2. This saw many new processes introduced to the forming of bricks and advances in kiln technology – all of which made brick both more readily available and also allowed mass production of special shaped bricks, different coloured bricks and high-quality glazed and facing bricks. This led to a golden age of decorative brickwork between 1860 and 1890, which saw decorative elements incorporated into even the most utilitarian of buildings. This reached its apogee with the construction of Templeton's Carpet Factory, Glasgow in 1892. Gauged brickwork continued to be used as a decorative embellishment throughout the 19th century, but saw competition in the form of preformed arch sets in the second half of the 19th century.

Technical features of note in this time are considerable. Recognisable bond patterns can be seen throughout British brickwork. The gauge to which brickwork was laid and the size of bricks both generally increased in the latter part of the 19th century. Both bond and gauge saw considerable regional variation throughout Britain. This is discussed more fully in Chapter 3 and saw, for example, the development of specific regional gauges in the North of England and Scotland. The joints between bricks generally reduced in height and width as bricks became more regular in size and shape. The 19th century also saw cavity wall brickwork come into use. Yet this is not always evident through the use of stretcher bond. Where headers are used to tie across the cavity, bonds such as rat-trap bond may be seen – as in the brick housing in Main Street, Newtongrange dating to 1872. Brickwork was also used extensively to form fireproof flooring in industrial structures.

The 19th century saw a period of expansion and diversification in the use of brick. The material was used to construct urban, industrial and transport infrastructure to support a country and society seeing unprecedented expansion and change. Advances in manufacturing techniques saw brick production expand and the range of brick types available also develop considerably. This saw brick incorporate new decorative and technical features in what may be regarded as the golden age of traditional brick construction in the United Kingdom.

Although beyond the scope of this work, as the definition of traditional brickwork in the context of this book ends in 1919, brick continued to be used extensively following World War I. The loss of skilled craftspeople during the war had a lasting impact on craft skills, as did the common use of cavity walled and rendered brickwork, which precluded the use of many of the decorative techniques from the previous century. Many fine examples of 20th century brickwork may be found, however, including the Taylor Street Ventilation Station, Cheshire (begun in 1925) and Velarde's English Martyrs' Church, Wallasey (1953). The craft skills required to construct and repair traditional brickwork in this period have been kept alive by a small number of craftspeople. Today brick continues to be a widely used material in both construction and the repair of existing structures.

Brick is a material with a complex history of use in the United Kingdom. With early uses by the Romans primarily archaeological, except for some reuse of Roman material in later extant buildings, the period from the 12th century onwards saw brick emerge as a building material in Britain. Medieval uses of brick were principally in high-status buildings, a trend that would continue until the 17th century. The Georgian period saw brick used extensively

Fig. 5.10 Templeton's Carpet Factory, Glasgow is one of the most decorative examples of Victorian brickwork and encapsulates a golden age of decorative brickwork.

Fig. 5.11 The 19th century saw brick used to house industrial workers on a hitherto unprecedented scale.

Fig. 5.12 Railway construction led to a high demand for bricks to construct infra-structure as well as allowing brick manufacturers to transport the material far and wide.

Fig. 5.13 Brick was used to construct industrial buildings, including tall chimneys; this was the tallest chimney in the world when it was constructed in Glasgow in 1857.

throughout the country and across buildings of all status from humble to prestigious. The 19th century, with its many advances in brick manufacturing and transportation, saw brick become the dominant masonry material across much of the United Kingdom. With the expansion of use and widespread employment of decorative features in the years between 1860 and 1900, this was a particularly prolific period in brick construction.

Technical aspects of brickwork have changed over time, as have the uses to which the material was put. Bond patterns developed from irregular in the early period, through the common application of Flemish bond in the 18th century, to the dominance of English bond in the 19th century. The characteristics of bricks themselves also changed, especially in the second half of the 19th century when advances in manufacturing led to the use of glazed and polychromatic brickwork.

Developments in the use of brick, in terms of application, and technical and aesthetic characteristics, can be used to aid in the dating of brickwork and in ascertaining the heritage significance of that brickwork. This evolution in craft practice and use has left a heritage of brick buildings across many centuries that require careful understanding, survey and assessment to ensure they survive for future generations.

5.6 Case study 5: Flatfield Farmhouse, United Kingdom – assessment of heritage significance

The farmhouse at Flatfield Farm presents an interesting case study in assessing the heritage significance of a vernacular brick building. Ascertaining the age of the farmhouse involved documentary research and the study of early maps and purchase records for land in the area in the 18th century. A construction date of 1785 was ascertained through this research. The building itself is of an unusual form of construction. The walls are mass brickwork, one and a half feet in thickness. They are constructed of irregularly bonded brickwork. Although no recognisable bond pattern is used, bonding does take place, with headers and stretchers laid in such a way that joints are broken between courses. The brickwork is constructed off a rubble stone base course, with a further stone course at eaves level. Stone lintels with a brick relieving arch are constructed over window and door openings. The bricks are handmade and vary considerably in terms of degree of firing. It is probable that the bricks were manufactured on site and fired in a small clamp kiln, although this cannot be proven for certain.

Flatfield Farmhouse provides an intriguing case study in attributing and appreciating heritage significance. As one of the earliest brick-built houses in Scotland, it is of truly national significance to Scotland's built heritage. Fewer than ten earlier brick houses survive, making the establishment of a date for the construction of Flatfield of particular importance. The building also has significance because the brickwork was constructed in a similar manner to mass earth construction in the local area. It is clear from an examination of the technical characteristics of the brickwork at Flatfield that it has more in common with earth construction than with brickwork in other parts of the United Kingdom in the late 18th century, a period that saw fine Georgian Gauged brickwork constructed elsewhere in the United Kingdom. Flatfield Farmhouse is therefore a transitional building between earth building and brick building traditions, making it highly unusual and of considerable significance in the evolution of construction history in the United Kingdom.

Fig. 5.14 Flatfield Farmhouse provides an intriguing case study in assessing heritage significance. Although not on the scale of Hampton Court Palace, buildings such as this play an important part in the development of brickwork.

Fig. 5.15 Technical features such as irregular bond and a wall thickness that follows earth construction practices add to the heritage significance of the building. An appreciation of the part that materials and craft practice play in heritage significance is important when assessing traditional construction.

References and further reading

Ayres J. (1998), *Building the Georgian City*, New Haven, CT: Yale University Press.

Brunskill R.W. (2009), *Brick and Clay Building in Britain*, New Haven, CT: Yale University Press.

Campbell J.W.P. (2003), *Brick: A World History*, London: Gollancz.

Clifton-Taylor A. (1987), *The Pattern of English Buildings*, London: Faber.

Jenkins M. (2018), *The Scottish Brick Industry*, Catrine: Stenlake.

Jones E. (1986) *Industrial Architecture in Britain 1750–1939*, London: Batsford.

Lloyd N. (1925) *A History of English Brickwork,* London: Institute of Clayworkers.

Lynch G. (2012), "Tudor Brickwork", in *The Building Conservation Directory*, Tisbury: Cathedral Communications.

Warren J. (1999), *Conservation of Brick*, London: Butterworth-Heinemann.

Woodforde J. (1976), *Bricks to Build a Brick House*, London: Routledge.

6 Defects and traditional brickwork

In common with all construction materials, brickwork will decay if not maintained. The process by which fired clay bricks decay is a complex one. It can be induced by the actions of salt or biological growth, or through inappropriate intervention. Decay of brickwork most often occurs in the presence of excess moisture. A key part of identifying defects is finding any sources of excess moisture saturating the brickwork. This chapter discusses various decay mechanisms and gives an indication of their possible cause. Identifying defects in traditional brickwork is often a key aim of survey and assessment work.

6.1 Salts and brickwork

Salts can cause deterioration in brickwork. Salts within brickwork are generally seen by the presence of efflorescence, which presents as white streaks on the surface of bricks or, in more severe cases, a build-up of white crystals or a powdery deposit over larger areas of brickwork. Salts damage bricks because they are soluble and can dissolve and recrystallise, often within the pores of the brick at the point of evaporation. The force of this crystallisation can be strong enough to force the internal structure of a brick apart, leading to damage and deterioration of the face brickwork. Efflorescence is therefore a potential cause of spalling, as discussed in section 6.2. Efflorescence is the outward sign that salts are being mobilised within brickwork. Problems with salts are further compounded by the fact that some are hygroscopic, attracting moisture from their surroundings.

Salts can enter bricks through capillary action from rainwater, groundwater or condensation, although more commonly water ingress brings to the surface salts already present in bricks or mortar. This is particularly common in brick chimney flues, where salts and other contaminants from smoke and flue gasses have penetrated the brickwork over a long period. When these react with dampness, either from condensation within the flue or from rainwater

DOI: 10.1201/9781003094166-7

penetration and become soluble, they can enter the brick and result in deterioration. This is most likely to occur where a redundant chimney has been sealed without providing adequate ventilation. In marine environments, salts can come from the sea as shown in Figure 6.1 and in winter, road salt is a further threat. Where brickwork has been cleaned, it is possible that the cleaning agents may have introduced salts if not rinsed or handled correctly, a further deleterious effect of cleaning to be watchful for. The most common source of soluble salts in brickwork, however, is through the use of inappropriate Portland cement-based mortar for repairs over original lime-based mortar.

When surveying a traditionally constructed brick building, it is often only required to note the presence of efflorescence and the possible source. Efflorescence is the most commonly seen outward evidence of the presence of salts within brickwork. Salts in bricks can cause damage in many different ways. The crystallisation of the salts and the cycling of dissolution and crystallisation within pores in clay bricks can lead to considerable stress within the brick itself and lead to the decay of that brick. Salts can also cause damage by accelerating corrosion of iron work embedded within brick walls. However, while certainly affecting the aesthetic appearance of brickwork, efflorescence is in some cases not damaging to the brickwork, so a careful assessment should be carried out regarding the extent to which salt is damaging brickwork.

When salt crystallisation occurs beneath the surface of bricks rather than on the outer surface (a process known as cryptoflorescence or subflorescence), the likelihood of damage occurring is higher. If it is felt that salts are causing damage to brickwork, it is possible to take samples of the efflorescence and subject them to laboratory analysis, as discussed in section 7.4. This will help diagnose which salts are present and will also help to identify the source of these salts. Some salts are potentially more harmful to brickwork than others – sodium sulphate, for example, is regarded as harmful to brickwork. The degree to which salts pose a problem to brickwork will depend on the extent of the problem, the type of salt present and the vulnerability of the bricks themselves. As discussed in section 3.2, different brick types have different characteristics, which will influence how they are affected by salts.

As well as identifying the possible source of salts affecting brickwork, such as cement pointing, contaminates from chimney flues or road salt, the source of any excess moisture that is mobilising the salts should also be identified. This is likely to come from an underlying building defect such as blocked gutter or downpipe, or a defect within a chimney stack. Where salt is present on the surface of brickwork in the form of efflorescence, or it is suspected as a cause of spalling or other surface decay, this should be recorded in survey work and potentially investigated further if required. Salts are both a cause of decay and deterioration in brickwork and a symptom of wider building defects.

Fig. 6.1 Efflorescence affecting both brick and stone masonry in a railway bridge; the source of the salts is likely to be large areas of Portland Cement with water penetration from the bridge deck above.

Fig. 6.2 Efflorescence in mortar joints, likely migrating from contaminated ground water behind this retaining wall. As the joints are formed of lime mortar, the moisture and therefore the salts are largely dissipating through the mortar joints.

Fig. 6.3 Subflorescence showing as a white build-up on the surface of brickwork; again, cement mortar is the likely source. The soft brick is beginning to erode, as seen in the top left header brick (copyright HES).

6.2 Spalling and surface decay

The most common form of deterioration affecting the fabric of brickwork is a loss of material from the outer face of bricks, generally referred to as spalling. Spalling can be caused by a number of decay mechanisms. The most common cause is water penetrating the brick and expansion and contraction of this water through the freeze–thaw cycles in winter exerting pressure on the brick face, resulting in it being "blown off", or spalled. There will often be an associated building defect that results in excess moisture saturating the brickwork. As discussed above, salts can also cause spalling as their expansion can break apart the internal structure of the brick. Accumulated atmospheric pollution can form a surface layer, which may detach from the surface of brickwork, removing the surface of the brick with it. A far more common situation where this occurs, however, is in the use of cement render, as discussed in section 4.6. The outer surface of bricks can become detached along with cement render if it fails.

The use of hard, cementitious mortar as a repointing mortar over original, softer, lime mortar can result in increased levels of spalling as it prevents moisture dissipating from the brick through the mortar joint. The mortar should act sacrificially and allow water to escape from the brickwork; mortar should not be harder than the brick it beds for this reason. Another possible cause of spalling, if it occurs over a widespread area, is defects within bricks themselves. This could be due to under-firing, as a result of which they were too porous for the application to which they were put, or there could simply be an isolated brick of lesser durability. This is less likely to be the case with traditionally constructed brickwork, as it is likely to have been in situ for in excess of 100 years. Where widespread spalling is seen on traditional brickwork, the cause is more likely to be the use of cementitious repointing or a situation as seen in Figure 6.17. where an internal wall has become an external wall. If bricks have been cleaned using abrasive methods, this can also lead to spalling as the bricks, lacking the tougher outer fire skin, are more susceptible to water penetration and subsequent spalling.

Whatever the cause of spalling, once the "fire skin" formed during firing is lost from a brick, there is a higher probability of needing replacement. The degree of surface loss of a brick that has suffered spalling can be ascertained by measuring the depth to which the brick surface has been lost when compared to a neighbouring sound brick. For example, the brick shown in Figure 6.4 has lost between 30-40 mm of its surface compared with the sound brick in the course below at varying points. Simply ascertaining the degree of surface loss, however, is not necessarily an accurate indicator of the vulnerability of the brick to further deterioration. A dense brick of relatively even consistency is likely to be able to tolerate some loss from the face of the brick without suffering further deterioration (once the cause of this deterioration is rectified) than a brick with a relatively thin durable surface that, when lost, exposes a softer interior. It is this vulnerability to further decay and deterioration that influences decisions about whether a brick needs to be replaced; this requires careful consideration during survey and assessment.

Different types of brick suffer spalling in different ways. Where a brick is of a type that is dense, durable and of a reasonably uniform consistency throughout, it will be less susceptible to further spalling, even if the front face is lost. For example, a pressed facing brick or engineering brick is of fairly uniform consistency. Even where some degree of surface loss has been suffered, such bricks may be able to perform satisfactorily in situ. However, bricks such as those manufactured from colliery shale, while tough and durable on their outer face, can be softer on the inside, as seen Figure 6.7. Such bricks, having lost their outer surface, are unlikely to be durable in the long term and are likely to require

Fig. 6.4 The spalling in these bricks is considerable and the interior of the brick clearly vulnerable to further decay; these bricks are likely to need to be replaced.

Fig. 6.5 The yellow special bricks forming the polychromatic string course have suffered extensive spalling and are likely to need to be replaced; the main facing bricks are in better condition.

Fig. 6.6 The outer fire skin of these bricks has been lost as a result of abrasive cleaning.

replacement. Handmade bricks can vary considerably in terms of durability and are likely to require individual assessment when found to have suffered spalling.

6.3 Defects within bricks

There is a tendency to regard traditional building materials such as bricks as either infallible or exceptionally deficient in performance. In reality, the performance of traditional building materials falls somewhere between these two extremes. While historic bricks have, by definition, survived in situ in excess of a hundred years in many cases, there can still be deficiencies within historic bricks that will become apparent, especially where a building is suffering from a lack of maintenance. In some cases, bricks will fail in a seemingly random pattern over the face of a wall. This can sometimes be attributed to deficiencies within the bricks themselves. Being a manufactured product derived from a naturally occurring material, as discussed in section 2.1, bricks can be of differing quality. Some may have inclusions such as small stones, as seen in Figure 6.8, or particles of quicklime. If the quicklime becomes saturated over a long period, this can result in "lime blowing", where the fired brick takes in moisture; the lime particles then slake and expand, and burst the surface from the brick, causing spalling. Likewise, where there are large inclusions, the brick can be vulnerable to frost damage. Some bricks are also under-fired, and these are generally softer, more porous and therefore more vulnerable to decay. Where bricks are seen to fail in a random pattern, this may be attributable to deficiencies within the bricks themselves.

The fire skin of a brick may be defined as the dense surface layer on an unglazed fireclay brick. The composition of the clay and the temperature and duration of firing will determine the depth, degree of vitrification and porosity of the fire skin. In most traditional bricks, a fire skin is formed to a greater or lesser extent. An exception is the soft rubbing bricks used for gauged brickwork.

Defects may have emerged within bricks soon after construction, but may be stable for the present. If the defect has been present since manufacture, and a brick has been in situ in a wall for over a century, it is likely that this defect will not pose any significant risk in the future. These defects may also have arisen as a result of a cause that has been rectified – for example, a blocked gutter that has been repaired. If defects within bricks are found to be stable and not deteriorating further, as was found to be the case in the brick shown in fig. 6.10, it may be possible to leave these bricks in situ in a wall. It would, however, be advisable to draw attention to such defects so they may be monitored in future surveys. There are also a number of defects within brickwork that may be regarded as "live", those which are likely to deteriorate further and are likely to cause decay and damage in the future.

Given the prevalence of cleaning brickwork, particularly in the 1980s and 1990s, when traditional brick buildings are being surveyed care should be taken to note where any cleaning has taken place in the past. Many brick cleaning methods can cause damage to brickwork, particularly abrasive methods such as sand or grit blasting. These methods of cleaning can scour the outer face from bricks. Chemical cleaning can have a similar effect, and can also introduce harmful chemical residues and salt into brickwork. Where it is thought that brick cleaning has caused damage to brickwork, it may be possible to access records relating to a building to identify the method of cleaning that was undertaken. Cleaning is therefore a further potential cause of damage and decay in traditional brick buildings.

Care is required not to mistake the natural surface finish of some bricks with those that have suffered decay. Bricks manufactured from colliery shale (often termed colliery or common bricks) present an open-textured finish that could be mistaken for defects or decay.

Where small holes are found in bricks, there are a number of possible explanations. Where a garden is adjacent to the brickwork, it is possible that nails have been fixed into the bricks in order to train plants to grow. Where bricks are soft, as would be the case with rubbing bricks used for gauged brickwork, holes may be caused by masonry bees burrowing into the bricks. Holes may also be a product of defects within the bricks themselves, as described above.

An individual brick, or indeed a small area of localised brick decay, is unlikely to affect the overall structural stability of a brick structure. The exception to this would be where relatively thin brickwork is used in a way that supports significant loading. An example of this would be where a jack arch brick floor was used, as can be seen in what are termed "fire-proof mills" discussed in section 3.7.8. A further exception would be a situation where an arch is constructed of brick if individual voussoirs have failed. This can affect the overall structural stability of the arch, especially if there is not a substantial lintel above. In most masonry walls, small areas of localised decay will not affect overall structural stability. Ascertaining the structural purpose of a piece of brickwork where decay has taken place is therefore crucial to evaluating its severity.

Assessing the severity of decay is particularly significant ahead of proposed repair work or rebuilding. Where areas of brickwork are being repaired, it will not always be necessary to replace all bricks affected in an area of decay. In some cases, it will be possible to cut out and reuse original bricks in conjunction with replacement of those that are badly decayed. This will require a careful assessment of the condition of the bricks and equally careful removal. This strays into the realm of repair specification, however, which is beyond the scope of this book.

Fig. 6.7 The interior of this brick has been fired to a lower temperature than the outer fire skin, forming what is sometimes termed a "black heart brick"; once the fire skin is lost, such bricks often lack durability.

When presenting the results of a survey of traditionally constructed brickwork, any spalling or loss of surface materials should be highlighted. It may be appropriate, depending upon the level and detail of the survey and its purpose, to make comment on the possible need to replace bricks or indeed to note that, despite spalling, it is possible to retain the bricks in situ. As noted above, however, this requires a thorough understanding of the bricks themselves and the potential for further decay and deterioration. It should not be assumed that a loss of surface material from a brick automatically precludes its retention in a wall. It is, however, certainly a cause of concern and something that

Fig. 6.8 A loss of the surface of this brick has revealed large inclusions in the clay in the form of a pebble and coarse aggregate; if left in situ, an assessment of the durability of the brick would be required.

Fig. 6.9 This extruded brick has lost such a significant amount of its surface that the perforations formed during manufacturing are exposed; this brick is unlikely to be durable in the long term if left in situ.

Fig. 6.10 The small declivities in these bricks are likely to have formed as part of the manufacturing process; as they are not causing any associated decay, they are not a cause for concern.

warrants further monitoring and investigation. The cause of the spalling should also be investigated and noted. Where appropriate, this may require repairs to the wider structure to deal with other defects, such as rainwater goods or roof coverings.

6.4 Failure of timber, metal and stone

Despite the focus of this work on bricks and mortar, no material in a traditional building exists in isolation. Defects seen in brickwork can be a result of interaction with other materials. This is most commonly seen where ferrous metal or timber is built into brickwork and these materials have suffered decay. Stone is a third material often used in conjunction with brickwork, which can lead to defects.

6.4.1 Failure of reinforcement

Timber was embedded into brickwork in traditional construction for many different purposes. Bond and chain timbers provided longitudinal reinforcement to brickwork during construction. Timber is used for lintels and was also built into brickwork to provide the means to fix building elements such as windows, staircases and internal wooden finishes. Floor joists were often built directly into pockets within brickwork. Where timber is embedded in brickwork for any of these purposes, if it remains dry there should be little risk of decay. However, when timber is exposed to excessive moisture and lacks the ability to dry out, various forms of timber decay may establish themselves, including wet rot, dry rot and insect infestation. Where timber decays, this can have a serious effect on the stability of brickwork, potentially leading to structural movement, cracks and instability. Where structural issues, such as cracks, bulging or other forms of distortion, are found in survey work, decay of built-in timbers is a possible cause. This will require further investigation, which may necessitate the use of more invasive survey methods. Where invasive or destructive investigative methods are being considered, statutory consents and considerations around heritage significance are likely to apply. Decay of timber is therefore a further potential cause of failure within brickwork.

As discussed in section 3.6, ferrous metal – most commonly iron – may also be encountered in conjunction with brickwork, particularly that constructed in the 19th century. The most common application of iron within brickwork is cavity wall ties, which may be found in buildings from as early as the 1860s. Lengths of hoop iron were sometimes built into brickwork to fulfil the same purposes as bond timbers. Wrought iron was also employed to tie elements of brickwork together, such as jack arch floor construction. Metal may also be found forming fixing points for doors or as structural members behind brickwork, as discussed in Case study 6.

Ferrous metal that is suffering corrosion can cause damage to brickwork. As iron corrodes, it expands considerably in volume. This can exert significant force within brickwork, which can exceed the sheer and tensile strength of that brickwork through a process referred to as oxide jacking. This can create cracks in brickwork; it can also damage mortar joints and cause the spalling of bricks. These defects brought about by embedded metal in brickwork can cause serious problems if not corrected. Where it is suspected that ferrous metal corrosion is leading to the decay of brickwork, as with timber, this may require further invasive investigation.

The root cause of decay of both timber and metal in brickwork is moisture. If corroded metal or decayed timber reinforcement is found during survey work, the source of the moisture that is causing the corrosion should be identified. This could be a building defect but could also be interstitial condensation. This is more likely to be the explanation if impermeable insulation has been applied to brickwork, altering the dew point within the structure – especially where impermeable insulation has been applied internally. The effects of insulation on the moisture content of masonry are a complex issue; further information on this subject can be found in the References and Further Reading section at the end of this chapter.

6.4.2 Stone decay and brickwork

Where stone is used in conjunction with brickwork, this can also lead to a variety of defects. This most commonly takes the form of stone and brick elements delaminating from each other. This may be a result of structural movement or be due to differential performance in relation to a variety of factors, including thermal movement. Where a stone elevation links to one of brick, it is important to ensure that the two remain fully bonded together. This

Fig. 6.11 Bond timbers built into 18th century brickwork; this section of brickwork was opened up to investigate the possibility of decay, which was found to have taken hold of the timber.

Fig. 6.12 Movement between the brickwork and stone at quoins; further monitoring may be required to ascertain whether this is a serious structural issue.

Fig. 6.13 A failure of the stone cope at this gable is leading to water saturating the brickwork below; stone elements are often used at vulnerable points in brick structures.

can be an area where original construction has become defective and additional reinforcement measures are required. Where there is any evidence of movement between brick and stone elements of buildings, this should be noted as a potentially serious defect.

Stone was used in conjunction with brickwork for a variety of purposes. In some cases, these were decorative and include features around window reveals, string courses and cornices. More commonly, stone was employed for technical purposes. This includes the use of stone for window sills and lintels. As discussed in section 3.7.3, stone was employed as a façade material with a brick backing. This most commonly takes the form of brickwork used as a backing material to ashlar masonry. In the context of garden walls, brick is also sometimes used as a lining material to rubble stone walls. As bricks, being relatively small units of masonry, present a large number of joints at a wall head, stone copes can often be seen on exposed brick gables or structures such as garden walls. Where these are found, their use should be recorded. If stone copes are suffering from displacement, or have moved out of position, this could be a serious defect. Not only are the loose copes themselves a potential threat, but it is also likely that the defective copes will be allowing water to penetrate into the brickwork. The condition of stone copes should be assessed carefully, with their position, condition and any mortar joints that may need to be repointed highlighted in survey and assessment.

There are also a number of interactions between specific stone types and fired clay bricks that can be problematic. Limestone, for example, can result in salt entering into brickwork, with repeated cycles of dissolving and recrystallisation leading to the spalling and decay of the surface of bricks over time. Differences in the performance characteristics of brick and stone can also cause decay and damage. If stone is softer or more porous than the brickwork surrounding it, water may come to saturate the stone, leading to decay. The opposite can also occur where water runs off hard, impermeable stone types, saturating neighbouring brickwork. As with most defects affecting traditional brickwork, where problems of this type are encountered it is likely that a wider building defect or lack of maintenance is contributing to excess water concentrating on a particular part of the building fabric.

6.4.3 Defects in cavity walling

Despite cavity wall construction being more commonly associated with 20th century building techniques, as described in section 3.7.2, cavity walling has a history of use stretching back to the 1820s, with metal wall ties first being used in the 1860s. Those surveying and assessing traditionally constructed brickwork may therefore encounter cavity wall construction.

The most common defect found in brickwork constructed as part of a cavity wall is a failure of wall ties. This can manifest itself in various ways. Iron or steel wall ties will rust and turn to ferrous oxide if exposed to oxygen and moisture. This will occur most commonly in the section of metal embedded in the outer leaf, progressively spreading into the cavity between the two leaves, and in some cases, even into the inner leaf section. Age is a good correlation with severity of corrosion so it is more likely in very old cavity walls. Particularly porous masonry units and/or eroded mortar joints will accelerate the process.

Where wall ties corrode, they may expand to many times their original size, in some cases two to three times the original size of the tie, and cause oxide jacking within joints between bricks. This pressure caused by the expanding metal can cause brickwork to become deformed, bulge or, more commonly, cause horizontal cracking within bed joints of brickwork. Cracking may also occur in other patterns; where there is any cracking in a cavity wall, corroded ties should be considered as a potential cause. Where a cavity wall is exhibiting any form of structural movement or the characteristic cracking caused by defective wall ties, this should be recorded and further investigation should be carried out.

It is possible to use borescopes and micro camera systems to inspect the cavity between two brick walls. This can be particularly useful when trying to identify whether a cavity has been bridged. The bridging of cavities through the accumulation of rubble, decayed mortar or other detritus over time can cause moisture transfer from the outer to the inner brick walls. These defects, as well as design details such as defective copings across gables, flashings above the inner leaf, and masonry sills crossing the cavity, can contribute to corrosion of the tie sectioning the inner leaf, and cracks may then manifest through the plaster. The use of micro camera systems can detect the presence of such defects without resorting to more invasive investigative methods, although some skill and experience are required. Although it is beyond the scope of this book, the failure of cavity wall insulation is a further defect associated with cavity brick walls. Further guidance on this problem can be found in the References and Further Reading section at the end of this chapter.

Where cavity obstructions obscure a borescope view, both bed joints and perpend joints can be cut into to ascertain whether cavity ties have suffered decay. In some cases, it may be necessary to remove bricks to gain access to the cavity, particularly where an outer leaf is more than half a brick thick. Care should be taken when cutting into joints to prevent, as far as possible, debris falling into the cavity. Appropriate steps also need to be taken where cavity wall insulation is present to ensure this is not damaged.

Wall tie corrosion can result in structural failure in the outer leaf. However, this issue is not confined to metal ties; a further fault may be identified with cavity walling, where brick or stoneware ties have fractured. This may be caused by structural movement within a wall and effectively results in the creation of two thin half brick walls standing largely independent from each other. Without adequate tying to the inner leaf, the slenderness ratio of the outer leaf can quickly become critically high. In extreme cases, this can lead to the actions of wind and weather causing the untied brick walls to move or deform. The use of specialist investigative techniques for identifying the presence of cavity wall ties is discussed in section 7.7.

Fig. 6.14 Cavity walling, the outer leaf of which is a single brick thick. This can make identification of the wall as incorporating a cavity difficult, as it was constructed as a solid wall in Scottish Bond. Also note the use of common brick backing with pressed facing bricks on the outer façade.

Fig. 6.15 Decayed cavity wall ties removed from a wall.

6.5 Defects and deterioration in mortar

Throughout the life of a masonry structure, there will inevitably be some deterioration and loss of mortar from joints. This will, in turn, lead to a need to repoint areas of brickwork. If unaddressed, deterioration of mortar can result in water penetration, internal dampness, decay of timber and internal finishes, and ultimately structural instability in brickwork.

Deterioration of mortar over many decades is a natural part of the way traditional structures work. Moisture should have the ability to dissipate through mortar joints in traditional brick buildings, with the mortar acting sacrificially to the bricks. As moisture from brickwork dissipates through mortar joints, the actions of this moisture movement can result in deterioration of mortar over time. This can be exacerbated by the action of frost as freeze/thaw cycles can cause mortar to decay. This is not necessarily indicative of any defect with the building as the intention was always to have moisture dissipate through mortar joints.

There are several signs that mortar within the joints between bricks has deteriorated. In extreme cases, this will be visible as open mortar joints where the mortar has completely eroded from the face of the wall. Joints that are open with no mortar are clearly indicative of a need to repoint, and should be highlighted in any survey or assessment of traditional brickwork. Where mortar is found to be loose or crumbling (often categorised as friable), this is indicative of a likely need to rake out and repoint. It is also possible in some structures to discern that mortar has lost much of its strength by becoming loose from the joints, in which case it will be easily scraped out of the joint with a blunt tool. This is likely to be caused by the binder in the mortar degrading over time, but can also be a result of poor repairs with insufficient after-care for lime mortar.

The opposite problem may be noted where mortar is too strong for the brickwork it has been used to point. Where inappropriate mortar of too great a strength has been used, this will generally be a cement-based mortar. This can lead to deterioration of fired clay bricks rather than the mortar acting sacrificially.

When considering the need to repoint an area of brickwork during survey and assessment, the first stage is to assess exactly what parts of the building need repointing. In many cases, it will be unnecessary to repoint an entire brick structure or even an entire elevation. As some parts are more exposed than others, mortar will fail at different rates. Before any work is carried out, it is important to assess which parts of a wall are to be repointed and which are sound. As a general rule, providing the joints are not allowing moisture to penetrate, they do not need to be repointed until they have eroded to a depth as great as their height in the case of bed joints. There are always exceptions to every rule, however, and there may be situations where a mortar joint needs to be repointed, even where it has not eroded to a greater depth than its height. If it is felt that missing or eroded mortar is leading to decay or water ingress into brickwork, this should be noted as a potential future repair.

As is the case with removal of bricks, the decision to repoint mortar needs to be considered carefully. Heritage significance, aesthetic and technical considerations will all have to be balanced in any assessment of mortar. Mortar may be considered to be of heritage significance, especially where historic joint finishes are found – as can be seen in Case study 2. If survey work has identified historic jointing or pointing finishes, these should be recorded and highlighted where survey work is recommending repointing as necessary. Aesthetically, repointing can have a very significant impact on the appearance of traditional brickwork. If the mortar used in repointing is of the wrong colour, or if it is not applied using traditional craft practices, this can severely impact the aesthetic character of a piece of traditional brickwork. The superficial spreading of mortar over joints that did not require repointing is unlikely to be successful in the long term.

From a technical perspective, mortar is clearly of considerable significance. As discussed in section 6.10, inappropriate mortar can induce decay of bricks. Mortar joints do need to fulfil their function, which is to prevent moisture ingress through open or decayed joints into brickwork. Therefore, if joints are found to be open or if mortar is found to be decayed, this should be recorded and highlighted as a repair need. The balancing of technical, aesthetic and heritage significance is therefore crucial when assessing mortar joints in traditional brickwork. A conservation-based approach should be taken to this assessment process in the same way that bricks themselves are considered. Only if mortar joints are open, or if the mortar is friable and crumbling, should this be noted as requiring immediate repointing.

Fig. 6.16 Extensive loss of mortar can be seen in this brickwork, with all joints needing repointing.

Fig. 6.17 Cement pointing causing decay and spalling of bricks; here mortar is accelerating decay.

6.6 Arch defects

Arches are an integral part of many traditional brick structures. With all the technical features of arches in brickwork identified, as set out in section 3.5, any defects present can then be noted. The most common defect to affect brick arches is slipped or otherwise misaligned voussoirs (the individual bricks that form an arch). This is a relatively easy defect to identify in many instances if the slipped voussoir is on the outer face of the arch, as seen in Figure 6.18.

However, if one of the voussoirs that make up part of an internal course of an arch has slipped, this will be harder to identify. A thorough inspection of the intrados of an arch should be made during survey work where possible to ascertain whether there is any slippage in this area. Regardless of the arch profile employed, the intrados should show a smooth and consistent curvature throughout.

A more serious potential defect within arch construction is for several voussoirs to slip or become displaced from the surrounding brickwork. An arch can also deform from its original profile if other structural problems are affecting brickwork. This will be evident through the form of the arch deviating from its original form. Defects of this type are likely to be found in conjunction with cracks in surrounding brickwork, although this will not always be the case. Should any of these defects be found in survey work, this is indicative of potentially serious structural movement and will require specialist investigation.

Where an arch has become displaced, diagonal cracking will often be seen in the brickwork above. This can also be seen where a timber lintel behind an arch has decayed or failed. Timber lintels will decay if they are subject to

Fig. 6.18 Slipped voussoirs are affecting this camber arch; given the poor condition of the brickwork above this could be a result of wider instability in the building, further investigation is required.

Fig. 6.19 Although still performing well, some of the bricks forming the concentric header courses in this arch are starting to spall; over time, this could lead to the arch failing.

excessive moisture. Cast iron lintels may also corrode if they are exposed to moisture. Inappropriate attempts to repair defects with arches can result in increased deterioration, especially where cement has been used to fill cracks and "glue" slipped voussoirs back in place, resulting in further deterioration of the surrounding brickwork.

Defects affecting arches can occur as a result of the surrounding brickwork, which forms the abutment of the arch, being insufficient to resist the force exerted by the arch. This can occur due to alterations in a building – for example, the introduction of openings, alterations internally around floor structures or emerging defects connected to foundations and supporting masonry. Defects in arches of this type are likely to show themselves in cracking on either side of the arch, or in the brickwork above the arch itself. Where issues of this type are found, specialist investigation is likely to be required to ascertain the cause of structural instability and assess the condition of the structure.

6.7 Cracks in brickwork

Cracks are one of the most obvious indicators of structural defects within brickwork. They can also, understandably, be one of the most concerning defects when highlighted in an assessment of the condition of a building. For this reason, where cracks are found in brickwork, they need careful assessment to avoid either under- or over-estimating their importance and implications for the structure as a whole.

Cracks may be defined as being "live" – which means increasing in width, depth or length – through brickwork. This can be indicative of ongoing and developing movement within a structure. In traditional buildings, cracks can be historic. This may have taken place soon after the construction as a building settled, when the ground took up its weight. Cracks may also be related to past defects that have been rectified and are no longer causing further decay. Ascertaining whether a crack is active and part of an ongoing problem, or inactive and indicative of a defect that is no longer an issue, is of critical importance to proper assessment of defects in brick buildings.

It is also important to distinguish between fully developed (FD) and non-fully developed (NFD) cracks, FD cracks run from one plane to another, or intersect with another feature or crack, meaning that they cannot develop any further. NFD cracks have not reached a natural conclusion – typically another plane. They may continue to develop without any external influence in the same way that a sheet of paper may continue to tear easily.

If cracks are found in a traditional brick structure, it is likely that a period of monitoring will be required. Further investigative work may also be required into the wider structural condition of a building – for example, considering foundations, ground conditions, roof and floor structures. This will help to establish both the cause of the cracking and the rate at which the cracks are developing if this is the case. It is generally recommended that monitoring should last for at least a full calendar year to allow measurements to be taken throughout all seasons. However, a longer period of monitoring is likely to be required – at least 18 months for a relatively small building and longer for larger engineering structures. This will take into account fluctuations in weather and atmospheric conditions, helping to ascertain whether issues around thermal movement may be exacerbating cracks or wet weather is causing ground heave to crack brickwork. Longer periods of monitoring may be required to gain further detail around the behaviour of brickwork that is suffering from cracks. Conversely, if cracking is extensive and has appeared in a short space of time, a serious problem may have developed requiring immediate remedial action. There are some parts of a building where cracks are likely to be more problematic than others. For example, where cracks are seen in chimneys, the brickwork may be as thin as half a brick, meaning that decay induced by flue gases becomes more of a problem. Cracking here may lead to significant and potentially dangerous instability. The purpose of monitoring can be as much to demonstrate/reveal the ordinary behaviour of a building – and therefore be a source of comfort to the building user/ owner – as to inform a problem.

While monitoring and specialist investigation will often be required when investigating cracks and their cause, some general observations can be made around cracking in brickwork. Where cracks appear clean and to have sharp edges, this may be indicative of a relatively recent defect emerging. In brick structures, a crack is regarded as more serious if it runs through both bricks and mortar joints. The location, shape and orientation of cracks can all provide clues as to their possible cause and/or nature of movement. It is important to establish whether a crack has arisen because of a localised issue (e.g. corroding metal lintel) or is indicative of actual movement and, if so, whether that movement is local or global. For example, horizontal cracks – especially through joints – may be indicative of defects with cavity wall ties or other built-in metal expanding and corroding, or it may indicate settlement of that part of the wall below the crack. It may be discovered that a crack has been filled with mortar in a previous attempt at repair and the crack has opened again, which may be indicative of a problem that is continuing or may be a result of the natural inflexibility of cement mortar. The use of hard Ordinary Portland Cement-based mortars to fill cracks will often exacerbate the problem. Where this has occurred, it should be noted and highlighted. Where a crack is emerging above an arch above a door or window opening, this is likely to be a result of movement within the wall, causing spreading of the arch abutments. A possible cause of this is decay of a timber lintel behind the arch. Defects of this type can affect both the wall above an arch and the arch itself.

The footings and foundations of traditional brickwork will vary considerably, depending on the date of construction, the location and the type of building being surveyed. Such foundations and footings will often not be observable during a non-invasive survey. The need to investigate such features of a traditional brick structure generally only arises where structural movement is suspected as a defect. Where this is the case, an invasive survey is likely to be required, with specialist consideration given to the nature of the underlying ground and the condition of footings and foundations.

6.8 Bulging, distortion and structural defects

Cracking is often the most outwardly obvious symptom of structural defects in brickwork. Other signs of structural defects within brickwork include the bulging or bowing (either inward or outward) of masonry and individual bricks or courses of brickwork being misaligned from those surrounding them. Brickwork can also become distorted through structural movement, with courses of brickwork appearing out of level as seen in Figure 6.24. Evidence of past structural problems with brickwork can sometimes be found in the form of tie rods and pattress plates. Where these are found during survey work, they should be recorded and their condition assessed at both ends.

When surveying a brick building to ascertain structural defects, a thorough visual inspection should be carried out. This should include looking along the face of a wall as well as looking straight on, as this will help ascertain whether the wall is delaminating, bulging or otherwise suffering from distortion. Where bricks are displaced out of level or face alignment due to structural movement, particular note should be taken of any areas where this leads to bricks projecting from the face of the wall. This will lead to the creation of ledges on which water may collect, leading to saturation and possible decay of bricks. There are situations whereby quarter brick projections are created as part of a decorative scheme, as discussed in section 4.5, but these should easily be distinguishable from instances of structural movement.

Bulging of solid brickwork can be caused by expansion and contraction of bricks. This may be as a result of changes in moisture or variations in temperature. The effects of this are likely to be more pronounced where different materials are used within a wall – for example, where a glazed brick façade is applied to a backing of common brick, the common brick will absorb a greater amount of moisture than the glazed brick, potentially leading to differential performance and subsequent structural defects. This is more likely to occur where a building defect is present and is allowing moisture to saturate the brickwork over a long period of time.

Where a brick wall was originally intended to be internal, this too may absorb considerably more moisture than would normally be the case with exterior brick walls. This, in turn, can lead to structural instability and decay. Where a gable wall has a large number of flues, if these are suffering decay – given that the walls between flues were often only a half a brick thick – these too can experience structural issues. Finally, the action of intervention within

Fig. 6.20 A significant crack running through bricks and mortar such as this is likely to require specialist monitoring and investigation by an experienced professional.

a traditional brick structure can alter the way it performs. Where large and concentrated loads are placed on traditional brickwork, particularly where this exceeds their original purpose, structural movement may be caused.

Visual inspection will provide initial evidence for structural problems. Where these are found, they should be recorded and highlighted in survey and assessment. Structural defects may also require further investigation, and where more invasive investigation is required, the use of a borescope may be appropriate. This will require the drilling of holes into brickwork, which ideally should be done through mortar joints, which are more easily repaired than bricks. A borescope can provide information on the condition of the core of a brick wall and may be particularly useful to investigate whether delamination between elements of brickwork of the type seen in Figure 6.25 is occurring. Borescopes can also be helpful when investigating cavity brick walls, as discussed in section 6.4. It should be noted that this is an invasive process (unless the borescope is being inserted into an existing crack), so statutory consent may be required – particularly where the brickwork is of historical significance.

Fig. 6.21 The crack in this image has previously been repaired with cement mortar. This has been ineffective at addressing either the crack itself or its cause. Where situations like this are found, it is likely that the crack will need to be raked out and a proper repair effected.

Fig. 6.22 The projecting pilaster seen here is delaminating from the surrounding brickwork. This can be seen with the cement mortar repair at the junction of the pilaster and the brickwork. The cause of this is decaying ferrous metal behind the pilaster.

Fig. 6.23 This crack has previously been repaired using cement mortar with reinforcement being used to prevent further spread. Although not the most appropriate repair strategy, the crack does not appear to have widened since the repair was carried out.

A range of other investigative techniques can be employed if structural defects are suspected. Radar can be used to detect voids or metal that may be present in brickwork. This can include hoop iron reinforcement or cavity wall ties. Thermal imaging, as discussed in section 7.10, can also be used to look into structural issues as it will highlight colder areas of brickwork if used correctly. Ultrasound has also been employed to find voids within masonry. These techniques are likely to require specialist professionals, experienced not only in their use, but also in the interpretation of the results. These specialist surveying techniques can be helpful for investigating structural defects in traditional brick buildings, and are discussed further in Chapter 7.

Fig. 6.24 The building seen here has suffered structural movement. The string courses can be seen to be out of horizontal alignment. The extent to which this movement is historic will need to be ascertained before any intervention occurs.

Fig. 6.25 In this example, the front face of the brickwork has delaminated from its backing. This has led to an area of collapse that has exposed the softer interior bricks behind.

6.9 Atmospheric pollution and surface deposits

Brickwork can be affected by a range of materials forming deposits on its surface. This can include atmospheric pollutants or the deliberate use of coatings such as paint or water repellents. The use of paint and other traditional coatings on brickwork is discussed in section 4.6. Where these survive, they are unlikely to be causing damage to the brickwork. Differentiating between traditional coatings, which are likely to have considerable heritage significance where they survive, and damaging surface deposits may require specialist analysis and interpretation, as discussed in Case study 7.

Over time, atmospheric pollution can form surface deposits on brickwork. This can take the form of a thin layer of soiling on the surface of bricks and mortar. Although often symptomatic of atmospheric pollution penetrating deeper into bricks and mortar, the extent of this depends on the type of brick that is used in a wall. Atmospheric pollutants can eventually cause damage to brick by creating stress between the more fired outer part of the brick and the softer interior. Where surface deposits of atmospheric pollution are identified during survey, this should be noted as a potential cause of damage.

Where surface deposits are found on brickwork, this is often viewed as a justification for cleaning the brickwork; however, cleaning of traditional brickwork can cause significant and irreversible damage if the wrong methods are used. Where cleaning is proposed following any survey and assessment, as much information about the soiling or surface coating to be removed should be garnered as possible. It is also imperative that information related to the bricks themselves is correctly identified. Only a thorough understanding of the bricks themselves and the material the cleaning is trying to remove will be sufficient to allow the method of cleaning to be considered in a holistic way.

This is likely to require specialist assessment by those who have experience of a range of cleaning methodologies and are knowledgeable about traditional brickwork.

One of the most common reasons for cleaning brickwork is to remove graffiti. However, it is possible to cause greater damage to the brickwork by removing graffiti than had it simply been left. The damage to the amenity value of traditional brickwork through the presence of graffiti must also be considered, however. Where graffiti is noted during survey work, a thorough understanding of the bricks onto which the graffiti has been applied will be necessary before any decision is made regarding its safe removal. Attempts to use surface coatings to deter graffiti may trap moisture within bricks as some are impermeable to moisture.

Chemical consolidants and other surface applied treatments are a further surface deposit that may be found in survey and assessment work. The use of such materials to stabilise brickwork or to prevent moisture entering bricks is generally unsuccessful in the long term. While they can appear to be effective initially, there has not been enough research carried out on the possible long-term effect on brickwork of such treatments. For example, the consolidated brickwork may shift the problem to an adjacent area, as is often seen where patch repairs using cement mortar are found. Alternatively, it may perform differently to the surrounding masonry and result in accelerated decay in later years. Surface sealants can result in a similar situation by trapping and holding in moisture. There is also a likelihood of the surface treatment failing and peeling from traditional brickwork, as seen in the case study at the end of this chapter. The use of both consolidants and sealants, if found during survey and assessment, should be noted and considered a potential cause of decay.

Fig. 6.26 Graffiti can be a common justification for cleaning, but any graffiti removal should be considered carefully, with full knowledge of the brickwork, to avoid damage.

Fig. 6.27 Surface deposits on brickwork can affect the aesthetic of traditional brickwork, but their removal should be considered carefully to avoid damaging the
brick beneath.

6.10 Previous repairs and interventions

Regrettably, it is often the case that previous interventions to traditional brickwork are a cause or exacerbating factor in decay and deterioration. Inappropriate
repairs are generally a result of the wrong method or the wrong materials
(although the two often go together).

The use of modern cement as a binder for mortar in repointing work is
one of the most common repairs that can cause or exacerbate the decay of
brickwork. Modern cement repointing of this type in brickwork that is of
traditional construction can prevent moisture dissipating through mortar joints,
causing it to become concentrated in bricks. Cement renders have a similar
effect over a wider area and can trap considerable amounts of moisture within
brickwork, leading to decay of bricks and materials such as ferrous metal and
timber, which may be built into brickwork. A potential inappropriate intervention in some cases that is becoming increasingly common in brickwork is the
use of impermeable insulation, which is often retrofitted to the external face of
brick buildings. This can have a similar effect to cement renders by preventing
the dissipation of moisture through brickwork. If used internally, the use of
inappropriate insulation can cause interstitial condensation, which again can
lead to deterioration in the condition of brickwork.

Other interventions that may have a deleterious effect on traditional brickwork include some cleaning methods and the use of unsuitable replacement
bricks. Some methods of cleaning brickwork – particularly those that are abrasive, such as grit or sand blasting – can scour the outer face from a brick,
exposing its softer interior. This can leave the brick vulnerable to future decay.
As discussed in the previous section, cleaning should only be considered where

there is a clear need to do so and only using methods that do not harm the brickwork. Although beyond the scope of this work, guidance on the cleaning of brickwork is contained in the References and Further Reading section of this chapter. Where brickwork is found to have been cleaned, this may be a potential cause of decay. In the case of the building in Figure 6.28, cleaning is likely to have been a causal factor in the surface decay.

When surveying traditional brickwork that has been cleaned in the past, as much information as possible should be obtained regarding the method/s used. Records should be retained of work of this type. If records do not survive, however, it may be possible to ascertain the method of cleaning used by examining the effects that this has had on the brickwork. Chemical cleaning may leave bricks discoloured; abrasive cleaning methods such as sand or grit blasting are likely to have scoured the face from the bricks, leaving them scarred and pitted, and producing a textured surface more open to decay. Where damage to brickwork has been caused by cleaning, even if the original method used cannot

Fig. 6.28 The cleaning of brickwork by abrasive means is an intervention that has caused significant harm from the latter part of the 20th century onwards. The bricks and mortar in this fireproof floor have been damaged significantly by the use of abrasive cleaning.

be identified, this should be recorded in a survey, as it is likely to have left the bricks and mortar in a vulnerable condition.

The selection of bricks during previous repair work can also be harmful to traditional brickwork both technically and aesthetically. In terms of technical performance, bricks that are either harder with a lower water absorbency rate or softer with a higher water absorbency rate than those used originally may decay at a faster or slower rate than those that surround them. There is also the possibility of introducing salts into brickwork where areas of repair have been carried out. Aesthetically, the wrong choice of brick can have a considerable impact on the heritage significance and overall appearance of a traditional brick building. The aesthetic characteristics of brickwork have been set out in Chapter 4 and include colour, shape and surface texture. The wrong choice of brick during repair can significantly impact the character of traditional brickwork.

When surveying traditional brickwork, any previous interventions that are causing decay or that may be potentially harmful should be recorded. Care should be taken to ensure that these are fully noted as being a result of previous intervention. This will help inform future work. In some cases, it may be necessary to leave these inappropriate repairs in situ, as removing them will cause more harm than good. This decision will need to be taken on an individual basis, depending on the individual building. As is readily seen from this section,

Fig. 6.29 The rebuilt section of this wall has used bricks that are incompatible with the softer handmade original bricks. The new section is also bedded in impermeable, inflexible cement mortar. This means that the rebuilt section will perform differently in many ways from the surviving original, possibly leading to instability and decay in the future.

Fig. 6.30 It has already been noted that cement repointing is one of the most significant negative repair strategies affecting traditional brickwork. A further example can be seen in this image.

retaining good records of interventions can considerably aid future survey and assessment work; in this regard, simple BIM models of the type discussed in section 1.4.2 are effective tools.

6.11 Dampness and moisture

The majority of decay mechanisms that can cause traditional brickwork to deteriorate occur due to excessive levels of moisture within bricks and mortar. Water can activate salt within bricks and mortar, and can lead to the deterioration of both – particularly in the presence of freeze–thaw cycles. Water will also lead to the decay of timber built into brickwork and corrode metal, which can cause oxide jacking. Due to these deleterious effects, correctly identifying where moisture levels within bricks are high is a key aspect in the successful survey of a traditional brick structure. Various methods can be employed when trying to identify the presence of moisture within brickwork. It is important, however, to fully understand the dynamics whereby traditionally constructed brickwork allows water to be absorbed and desorbed. This process is often referred to as breathability.

Traditional bricks and mortar will generally be porous to an extent. This means that moisture, either as a liquid or a vapour, can enter into the fabric of traditional bricks and mortar then diffuse out again as conditions change. The movement of moisture within traditional brickwork is a complex process that is influenced by many different factors, such as ambient humidity levels, temperature both inside and outside a building, and the properties of the bricks and mortar themselves. Moisture can come from many different sources. Externally, it comes from atmospheric conditions and the weather in the form of wind-driven rain or snow. Internally, brickwork can be exposed to significant levels of moisture through the actions of habitation, such as cooking, bathing and drying clothes.

Bricks are, in most cases, relatively absorbent of moisture. It is this ability to absorb and release moisture that makes traditional brickwork "breathable". It is impossible to prevent moisture from entering into traditional brickwork to an extent, regardless of attempts that are often made to use surface coatings or render to achieve this. Maintaining the free movement of moisture throughout bricks and mortar will give the brickwork its greatest chance of long-term survival. When surveying brickwork, it is important to understand this mechanism as it informs many considerations related to decay. This is not to say that brickwork should be allowed to become saturated as a result of building defects, but rather that traditional brickwork, if well maintained and allowed to function as it was designed to, is well able to deal with what may be termed "normal" wetting and drying cycles.

When considering moisture in brickwork, the term "significant dampness" is sometimes used. This is defined most succinctly by Burkinshaw and Parrett (2003, p. 11), who state that, "Significant dampness exists where there is sufficient moisture in a material to cause problems for that material, adjacent materials, the building and adjacent building, the users of the building, the client or any relevant party." The threshold for significant dampness in brickwork is determined by the situation in which the brickwork is found and the purpose to which that brickwork is being put. For example, the threshold for significant dampness in a freestanding brick garden wall is likely to be higher than that for a brick wall in a domestic building. It is also important to remember that while a certain moisture content may not pose problems for the bricks or even the mortar, this could still pose problems for any inbuilt timbers or ferrous metal that may be embedded in the brickwork. It is for the building professional to ascertain the moisture content of brickwork and to then decide whether this constitutes significant dampness or is an acceptable moisture content.

A further source of the moisture that can cause deterioration in traditional brickwork is condensation. Condensation is the process whereby moisture transfers in state from a gas to a liquid. There are two principal types of condensation that can affect traditional brickwork: surface condensation, whereby moisture forms on the surface of brickwork; and interstitial condensation, which forms within solid brick masonry walls or between two parts of a structure. In traditional brickwork, interstitial condensation may form in the centre of a mass brick wall or in the cavity between two leaves of cavity brickwork. It is also possible for interstitial condensation to occur where a brick wall has been lined with stone on the front face. Condensation is a natural part of any traditional building; it only causes problems in situations where a wall becomes wet with condensation and cannot dry out. When looking for the source of moisture that may be causing defects in traditional brickwork, condensation is a possible source to consider.

Where moisture is causing decay or deterioration of traditional brickwork, the source should be identified. Various specialist investigative techniques can be employed to identify moisture levels within brickwork and aid in identifying their source. These are discussed in section 7.3.

6.12 Biological growth

Biological growth on brickwork may be subdivided into two broad categories: small surface growth such as moss and lichen; and larger plants such as ivy or small trees. Both types of growth can cause deterioration of brickwork. Biological growth of any type is also an indicator of the presence of excess moisture in brickwork, the source of which should be identified as part of any survey. Defects such as blocked gutters and downpipes can provide the moisture for plant growth to survive. Biological growth can therefore be both a symptom and a cause of defects within brickwork.

Where brickwork needs to be repointed and there are open joints or friable mortar, biological growth can establish itself. Cracks within brickwork can also provide an entry point for such growth. Brickwork that is in sound condition and free from excess moisture will find it difficult to support the establishment of biological growth. Having found a point to establish itself, larger biological growth can penetrate deep into brickwork. This, in turn, can cause significant damage, dislodging bricks and causing instability – a situation that could have been avoided had the growth been tackled at an early stage.

The biological growth itself may well be causing defects within brickwork. Despite often being perceived as part of the patina of a building, creepers such as ivy can cause considerable damage to brickwork. The vines of creeping plants such as ivy can penetrate deep into brickwork, dislodging bricks and creating structural instability. If this then collapses, as in Figure 6.32, considerable damage can be caused to the brickwork. This effect is even more pronounced where larger plants establish themselves. If left unchecked, brickwork can become cracked or damaged. When surveying a traditional brick building, any signs of biological growth should be highlighted for future repair. The building in Figure 6.31 forms a useful case study. It can be seen from this example that biological growth in the form of a small tree has established itself in a crack in the brickwork.

Surface-level biological growth can also cause damage to brickwork. Moss and lichen can hold moisture against the surface of bricks and mortar, increasing moisture ingress and causing decay. Nearby growth in the proximity of a traditional brick wall can also cause damage. Larger trees can disrupt the foundations of traditional brick walls. Plants can also put brickwork in the shade, leading to problems with brickwork drying out. Where biological growth occurs near traditional brickwork, it should be considered a potential cause of decay.

When biological growth is identified during the survey of a traditional brick building, it should be recorded and highlighted as both a current repair need and also an indicator of possible wider failure within the building. The growth

Fig. 6.31 The widespread damage caused by biological growth can be seen in this example; extensive displacement of bricks has occurred, showing why growth such as this should be removed before it becomes established.

Fig. 6.32 Plants such as ivy, while sometimes erroneously thought to be part of the patina of age, can in fact cause considerable harm to traditional brickwork. It can be seen here that bricks and mortar have been adversely affected by the ivy.

itself should be recorded and, where possible, the source of the moisture that is allowing the growth to thrive also identified. Any damage caused by the growth should also be highlighted. Although this will be most obvious in cases such as shown in Figure 6.31, surface growth should not be overlooked, and nor should potential problems resulting from growth to nearby brickwork.

6.13 Wider building defects

Many of the defects found in traditional brickwork are caused by excessive levels of moisture. It is crucial to the success of any future repair to brick-work that this is allied with repair of any wider defects that may have led to the water ingress that caused the original failure. There is little point in carrying out costly repairs to bricks and mortar if the source of water that is causing the damage remains to cause decay in the future. The source of this excess moisture will generally be an underlying defect in a building. This book does not aim to provide a comprehensive guide to defects within traditional buildings; many excellent works on this subject exist already and are referred to in the References and Further Reading section at the end of this chapter. It is, however, worth briefly considering wider building defects that can be the root cause of the decay and deterioration of traditionally constructed brickwork.

One of the most common sources of excess moisture affecting brickwork consists of blocked gutters, downpipes and surface drainage. Blockages at any point in the rainwater disposal system of a building can lead to water saturating masonry and causing decay. Defects with internal plumbing systems can also lead to saturation of masonry. The identification of these wider causes of the decay mechanisms affecting brickwork is vital to ensure that the brickwork is effectively repaired and maintained in the future. There should always be cognisance of wider failures, defects and problems with a building, not just with the bricks and mortar. The use of thermography and other specialist techniques discussed in Chapter 7 can aid in the identification of these wider building defects.

Brickwork on chimney stacks is also a vulnerable area that can suffer many defects leading to moisture ingress. Missing lead flashing, decayed or missing mortar fillets and missing or cracked chimney pots can all allow excess moisture to penetrate into chimneys and flues. If it is noted that high moisture levels are present within the brickwork of chimneys and flues, it may be advisable to conduct an analysis of any salts that may be present, as detailed in section 7.4. This will help establish whether salt from past combustion gases is migrating into brickwork and causing decay.

Climate change is likely to lead to more extensive wetting of traditional brickwork in the United Kingdom and globally. It is also likely to bring more extreme weather events. Flooding is also likely to become a more frequent occurrence. Climate change and the increased moisture that it may bring are

Fig. 6.33 A failure of internal plumbing has led to the saturation of brickwork and stone masonry in this building over a long period.

factors of which those working with traditional brickwork should be aware, as it is likely to exacerbate existing problems with brickwork where they exist, with brickwork suffering more extensively from decay mechanisms if a building is not maintained correctly.

Although beyond the scope of this work, following repairs to brickwork a new system of planned maintenance should be put in place to ensure that problems of a similar sort do not re-emerge in the future. It is only through these two actions that repairs will truly be worth the time and expense involved. There is thus a clear link between failures in building maintenance and the need for repair, and this should be fully appreciated before any work is embarked upon. It is also vital that repairs are considered carefully and only carried out where absolutely necessary to avoid unnecessary loss of original fabric, damage and cost, taking a conservation based approach at all times.

Fig. 6.34 Failure of rainwater goods and parapet gutters is responsible for the water ingress to the brickwork in this building. This has led to biological growth establishing itself, and over time will cause considerable and costly decay to the brickwork.

6.14 Case study 6: Dumbarton Central Station, United Kingdom – survey of defects and their causes

This case study considers detailed survey work that examined defects to brickwork at Dumbarton Central Station, Scotland. This was undertaken ahead of a proposed programme of repairs. A number of defects and their causes were identified through visual survey:

- spalled bricks
- cracks through brickwork (bricks and mortar joints)
- salt efflorescence
- failure of a surface coating.

The causes of these defects were also investigated, and can be summarised as:

- corrosion of ferrous metal behind brickwork
- damage caused by previous cleaning
- damage caused by cement repointing
- application of a previous surface coating
- salts from cement pointing and probable salts from de-icing treatment
- previous repairs and rebuilding work.

These defects are explored in more detail below. It is worth noting at this stage that the majority of the defects identified above were a result of previous interventions (cleaning, rebuilding, cement pointing and surface coatings). This emphasises the importance of a thorough understanding of how interventions will affect traditionally constructed brickwork and why this understanding is critical to the success of any survey of such brickwork.

Cracks through the brickwork were identified at a number of points. These ran diagonally through both bricks and mortar joints, generally through four or five courses of brickwork. In some cases, attempts had been made to patch these cracks with cement mortar. The cause of this cracking was identified as ferrous metal, which had been built into the brickwork. This was corroding due to moisture ingress, leading to oxide jacking.

The bricks themselves were also suffering from decay at a number of points. This took the form of a loss of surface material from the bricks with a number of individual bricks also affected by spalling. One cause of this decay was identified as cleaning, most likely around 20 years prior to the survey taking place. In addition, a surface coating that had been applied to the brickwork after the cleaning had taken place was found to be peeling away. Although analysis of this was not undertaken, it was identified during the survey as likely to be a form of water repellent or sealant. It is likely that this was applied to prevent further surface decay following the cleaning. Cement repointing was also causing failure within the bricks as it was concentrating moisture in them rather than allowing it to dissipate through mortar joints.

In some areas, the brickwork had been taken down and rebuilt during previous repairs. It was noted during the survey that the bricks used in this repair work were smaller than those used originally. In addition, these areas were built to a different gauge from the original brickwork, with smaller bricks and wider mortar joints. This adversely affected the aesthetic of the brickwork and impacted its heritage significance. During the survey of the brickwork at Dumbarton Central Station, it was noted that where areas were to be rebuilt in future, bricks matching the original characteristics should be used and areas rebuilt to the original gauge. In areas where dogleg bricks had been used in the original construction at obtuse quoins, these had been inappropriately repaired. Rather than utilising special bricks in repair work, bricks had been cut on each side of the dogleg, the resulting joint being formed in cement mortar. This had a significant impact on the character and aesthetic of the brickwork. It also diminished the strength of the brickwork at the quoin, and by introducing large amounts of impermeable cement mortar to the brickwork created conditions whereby moisture may get trapped within the bricks.

The brickwork was also suffering extensive salt efflorescence. It was concluded that the efflorescence was most likely a result of cement mortar, and possibly also the new bricks that had been used in previous repair work. De-icing salts were also considered to be a potential source of the salts affecting the brickwork. The efflorescence was not subjected to analysis, as it was felt that the overall program of repairs proposed would reduce the problems – although

Fig. 6.35 The previous repairs to this obtuse quoin can clearly be seen here. The failure to use dogleg bricks and to match these to the existing brickwork has both an aesthetic and a technical impact.

Fig. 6.36 In this image, the extent of salt efflorescence can clearly be seen. Cement repointing is a likely source of this, with de-icing salts also likely to be a contributing factor.

it was noted that should efflorescence problems continue, further investigation may be required.

Only one specialist investigative technique was employed in surveying Dumbarton Central Station. The vast majority of survey work involved visual identification of both the characteristics of the existing brickwork, defects and their cause. Mortar analysis was commissioned as part of the survey work to help inform future repairs. This is referred to in section 7.1. A lack of specialist investigative techniques should not necessarily be considered a deficiency. In many cases, brickwork can be thoroughly assessed using a standard, non-invasive visual survey. Identifying the goal and purpose of a survey at the outset allows a more informed choice of techniques to be employed when surveying traditional brickwork, as discussed in section 1.1.

The main focus of this case study is problems with the bricks and brickwork. As considered in section 6.13, however, the presence of wider building defects was also a significant factor in the decay of the brickwork. At a number of points, rainwater disposal systems had failed, leading to extensive saturation of brickwork. Defects with surface drainage, coupled with hard, impermeable surfaces on the station platform, meant water became concentrated at the base

Fig. 6.37 The lighter patches on this brickwork are a failed surface coating. As the brickwork was cleaned in the past, affecting its performance, this surface coating is
likely to be a response to that failure.

of the brickwork, contributing to the decay of bricks, mortar and the inbuilt
ferrous metal, leading to corrosion. These wider defects within the building
contributed to the excess water ingress, causing decay to bricks and brickwork.

This case study has illustrated a practical example of the defects and their
causes described within this chapter. These defects were extensive and required
careful assessment. This assessment ultimately informed considerable repair
work to the brickwork and the wider structure. Previous interventions were
an aggravating factor in the problems exhibited in this building. This case study
acts as a strong example of how brickwork can suffer decay and the approach
taken to interpreting, recording and understanding defects within traditionally
constructed brickwork.

References and further reading

Ambrose J. and Tripeny P. (2006), *Simplified Engineering for Architects and Builders*,
Oxford: Wiley.
Ashurst N. (1994), *Cleaning Historic Buildings: Volumes 1 and 2*, Shaftesbury: Donhead.
Ashurst N (2011), "Cleaning Brickwork and Terracotta", *Building Conservation Directory*,
Tisbury: Cathedral Communications.
Baer N.S., Fitz S. and Livingstone R.A. (1998), *Conservation of Historic Brick Structures*,
Shaftesbury: Donhead.

Beckman P. and Bowles R. (2004), *Structural Aspects of Building Conservation*, London: Butterworth-Heinemann.

Bidwell T.G. (1977), *The Conservation of Brick Buildings*, Windsor: Brick Development Association.

Bonshor R. and Bonshor L. (1995), *Cracking in Building*, Bracknell: BRE.

Brunskill R.W. (2009), *Brick and Clay Building in Britain*, New Haven, CT: Yale University Press.

Burkinshaw R. and Parrett M. (2003), *Diagnosing Damp*, London: RICS.

De Vekey RC (2000), *BRE Digest 329: Installing Wall Ties in Existing Construction*, Bracknell: BRE.

Dearne T. (1821), *Hints on an Improved Method in Building* London: Harding.

Dickinson P. and Thornton N. (2004), *Cracking and Movement in Buildings*, Coventry: RICS.

Douglas J. (1995), "Basic Diagnostic Chemical Tests for Building Surveyors", *Structural Survey*, vol. 13, no. 3, pp. 22–7.

Espinosa-Marzal R.M. and Scherer G.W. (2010), "Advances in Understanding Damage by Salt Crystallisation", *Accounts of Chemical Research*, vol. 43, no. 6, pp. 897–905.

Franke L. and Shuman, I. (1998), *Damage Atlas: Classification and Analysis of Damage Patterns Found in Brick Masonry*, Stuttgart: Fraunhofer IRB Verlag.

Goudie A.S. and Viles H.A. (1997), *Salt Weathering Hazards*, Chichester: John Wiley.

Richardson B. (2000), *Defects and Deterioration in Buildings*, London: Routledge.

RICS and Historic England (2018), *Investigation of Moisture and its Effects in Traditional Buildings*, London: RICS.

Robson P. (1999), *Structural Repair of Traditional Buildings*, Shaftesbury: Donhead.

Watt D. (2011), *Surveying Historic Buildings*, London: Routledge.

Weaver M.E. and Matero F.G. (1993), *Conserving Buildings: A Guide to Techniques and Materials*, New York: John Wiley.

7 Specialist investigative techniques

The survey techniques outlined in earlier chapters provide a strong basis for ascertaining the technical and decorative characteristics of traditional brickwork, and also for beginning the process of identifying defects and their cause. There are, however, always times when more specialist investigative techniques may be required. This could be as a result of being unable to establish the causes of defects, or, in some cases, it may be the only way to identify particular technical characteristics of a piece of brickwork. This chapter sets out a range of specialist investigative techniques that may be used in survey work to fill gaps in knowledge and understanding.

7.1 Mortar analysis

"Mortar analysis" is a term used to describe a range of analytical tests that can be performed on samples of mortar taken from brickwork. Specifying that "mortar analysis" is performed is a somewhat misleading statement, as there are many different analytical techniques that can be used to ascertain different characteristics of mortar. The techniques employed to analyse mortar depend largely on the information the testing seeks to ascertain and what characteristics of the mortar need to be defined. Broadly, mortar analysis will take place in order to identify the binder used, the type and grading of aggregate and information related to the original mix ratio.

The first stage in analysing a sample of mortar taken from brickwork is generally visual analysis, both in laboratory conditions and in situ. The visual characteristics of mortar that can be seen in situ are discussed in section 3.4. When visually inspecting mortar in laboratory conditions, this will generally see a sample of mortar examined under a microscope to assess colour, size and type of aggregate, and establish the presence of a range of lime inclusions (under-burnt, over-burnt, as well as fully calcined and slaked lime lumps, unburnt lime or other inclusions such as shell fragments or hair fibres). Further information may be gained regarding these inclusions by later petrographic analysis. The condition of the mortar can also be described during such a visual analysis – for example, the sample may be described as friable or coherent. Hand specimen

DOI: 10.1201/9781003094166-8

microscope analysis may also provide information on weathering characteristics, surface staining and biological growth at this stage.

A variety of droplet tests can also be employed when analysing mortar in laboratory conditions. To take one example, this may see a drop of phenolphthalein used to test whether mortar has fully carbonated. A simple water droplet test can also be used to draw conclusions around the pore structure of the mortar and to help ascertain whether a water repellent has been used (this is discussed in the context of bricks in Case study 7). Other important droplet tests include testing with barium chloride and silver nitrate to detect the presence of sulphate and chloride, respectively, and the presence of organic compounds using hydrogen peroxide solution – which reacts with catalase (organic enzyme) to produce a fizzing reaction and release of hydrogen gas and water.

More detailed information regarding the composition of mortar can be obtained from other testing, with one of the most common methods employed being wet chemical analysis. Where this is carried out, a sample of mortar is weighed, dried, lightly crushed and then dissolved in an acidic solution. This dissolves the binder in the mortar, allowing conclusions to be drawn regarding aggregate used and mix ratio. There are, however, limitations to what can be ascertained through such wet chemical analysis, as acid will dissolve aggregate that contains calcium carbonate such as shell or limestone, and also any lime inclusions. Soluble silica can also be tested using wet chemical analysis in an effort to ascertain the presence of hydraulic materials. Mortar analysis can also be used to establish whether mortar contains OPC as a binder or additive if this is a potential cause of deterioration in bricks and to confirm what is observed by visual inspection.

Aggregate separation is a third type of analysis that can provide details of the type of aggregate used and its grading (size of aggregate and the distribution of different sizes of aggregate), helping to inform the type of aggregate used in future repair work. Aggregate separation and grading generally follow chemical analysis, as binder dissolution in acidic solution is the best method to separate out the aggregate and binder phases. Petrographic analysis, which sees thin sections of mortar examined under a microscope, can help to identify mortar properties such as pore structure, aggregate, binder and aspects of mortar preparation.

A range of other specialised tests can be undertaken when seeking to obtain information about mortar used in brickwork:

- *Scanning electron microscopy (SEM)* can show details of the microstructure of a mortar sample. It can also help identify the method of forming the mortar for example hot-mixed, putty lime or dry hydrate.
- *Mercury intrusion porosimetry (MIP)* tests will give an indication of the porosity and pore size distribution in mortar. The measurement of pore size distribution is an important factor/property that can be used to better understand the long-term behaviour and performance of the mortar.
- *X-ray diffraction (XRD)* can be used to identify crystalline components in mortar. This can help establish the binder used (hydraulic/non-hydraulic lime or ordinary Portland Cement), as well as the aggregate mineralogy and

any other crystalline components – that is, deleterious phases, salts and the addition of pozzolans and gypsum.

- *Gas chromatography* can be used to identify organic additives.

The type of analysis undertaken will depend on several factors, including the purpose of the analysis, the scale and budget of a project and the historical significance of the structure. Care must be taken, however, in interpreting results from mortar analysis to ensure that inaccurate conclusions are not drawn. A mortar

Fig. 7.1 A sample of mortar is being tested here with phenolphthalein indicator; this can help reveal various properties of mortar taken from brickwork (copyright HES).

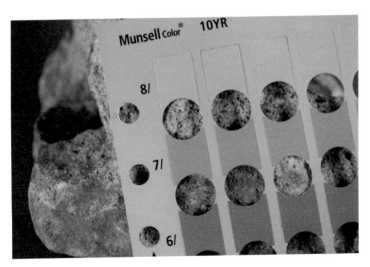

Fig. 7.2 This sample of mortar is being colour matched; the use of a Munsell chart can help standardise interpretation of colour of both mortar and bricks, although no dedicated chart for bricks is currently available (copyright HES).

sample taken from one section of brickwork may not be indicative of the mortar used in other parts of a structure. The mortar sample should be representative of the mortar used in the structure for its particular function (render/pointing/bedding/core), be relatively intact and contain representative aggregate and mineralogical phases (lime inclusions/remnant fuel/shell and so on. For petrographic analysis, for example, it is important that the sample is prepared through a representative section (i.e. a cross-section through the depth of a surface coating) and if information on lime source/manufacture/mortar preparation is required, then the thin section should be prepared through a section containing an area of representative lime inclusions. It is important to recognise that mortar analysis does not provide a ready-made specification for repointing work, but it can provide valuable information to inform such a specification. Mortar analysis can be a useful part of surveying traditionally constructed brickwork, but the results of such analysis require careful interpretation.

Fig. 7.3 This sample is currently reacting with hydrochloric acid during mortar analysis work to facilitate binder dissolution (copyright HES).

Fig. 7.4 The residual aggregate from Figure 7.3 ready to be graded (copyright HES).

7.2 Case study 7: Mortar analysis at Dumbarton Central Station

With thanks to Roz Artis at the Scottish Lime Centre for providing the report on which this case study is based.

This short case study discusses mortar analysis work as part of survey work ahead of repairs at Dumbarton Central Station; other aspects of the survey are discussed in Case study 6. The mortar sample was taken from the brickwork of the ticket office at Dumbarton Central Station, built in 1895. The sample taken weighed 59.99 grams in total, with the largest intact piece of mortar being 31.85 mm x 25.73 mm x 14.02 mm. This was clearly labelled and the location from which it was taken recorded on a site plan.

The sample of mortar was analysed using a variety of techniques in a laboratory situation. The sample was subjected to X-ray Powder Diffraction (generally abbreviated to XRD) to ascertain the crystalline components present and the binder used. It was found by means of XRD analysis along with observations from handheld microscopy and petrographic thin section that the mortar was mixed from a feebly hydraulic lime, most likely prepared as a "dry hydrate". In order to confirm the binder type, it was noted that further analysis by petrography would be required. The results of the XRD analysis were as shown in Table 7.1.

A number of other tests as described above were used to further analyse the mortar sample. A phenolphthalein indicator test confirmed that the mortar was fully carbonated and a water droplet test confirmed that the mortar was porous, as the droplets were rapidly absorbed and the water diffused throughout the mortar.

The addition of 10 per cent hydrochloric acid to the sample resulted in dissolution of the binder, enabling relative proportions of lime (and gypsum) to aggregate to be determined; where appropriate, proportions of insoluble binder were determined and factored into this calculation. Subsequent aggregate

Table 7.1 XRD analysis results – mortar sample from Dumbarton Central Station

Mineral	Notes	Percentage as per XRD analysis
Calcite ($CaCO_3$)	Calcium carbonate, carbonated binder from lime type binders	77.4
Quartz (SiO_2)	dominant component of the aggregate in the mortar	15.7
Brownmillerite	Calcium aluminium iron oxide, hydraulic lime or cement component	1.5
Greenalite	Greenalite, aggregate component	2.1
Anorthite		2.5
Cristobalite		0.3
Lazurite		0.6

characterisation was undertaken by means of dry sieve analysis and micro-scopic examination. The mix ratio was found to be 1 part feebly hydraulic lime hydrate to 1.01 parts sand/aggregate (by volume). It should be noted that the recommended mortar mix to replicate that which was found in situ was "1 part NHL 2 to between 2 and 2.5 parts washed building sand", demonstrating the principle described above that the results of mortar analysis do not directly translate across to a repair mix for brickwork in many cases.

The aggregate isolated from this sample was found to be moderately well graded, with aggregate retained from sieve size 1 mm and lower, with the highest percentage of grains being retained in sieve mesh size 0.125 mm with 40.3 per cent being of this size. Well-weathered lithic fragments and sub-angular quartz made up the coarser fractions. Smaller fractions were composed predominately of lithic fragments and angular to sub-angular quartz grains, and the buff/orange/cream tinted grains gave the sand its overall pale-brown colour.

This brief case study provides a good illustrative example of some of the most common tests used in mortar analysis for traditionally constructed brick-work. The exact testing carried out will vary depending on the aim of the analysis, but those tests set out here give those surveying brickwork a broad understanding of what can be achieved.

7.3 Measurement of moisture levels within brickwork

Many specialist investigative techniques can be used to identify moisture levels within traditionally constructed brickwork. These are concisely described below; recourse may also be made to the guidance listed in the References and Further Reading section at the end of this chapter, and in particular to RICS and Historic England (2018), which sets out methodologies that may be applic-able to traditional brickwork.

The clearest indicator of the presence of excess moisture within brickwork is the defects that this is likely to cause. Where salts are migrating to the surface of brickwork and causing efflorescence, or there is excessive biological growth in the form of plants, mosses or lichens, this is indicative of water ingress causing decay. While the technology described in this section may be helpful to con-firm this or provide more detailed mapping of the extent of the problem, the visual inspection of brickwork is likely to be the most cost-effective means of ascertaining whether moisture is a problem. The following techniques are also helpful.

- *Resistivity meters:* These are commonly used to determine the presence of moisture within brickwork. They work by measuring the reduced resist-ance in an electrical current between two electrodes. However, resistivity meters are most commonly calibrated to be used in measuring moisture in timber, rather than masonry, making their use in measuring moisture in brickwork uncertain in some cases. It has been found that resistivity meters can over-estimate moisture in brickwork where salts are present, for example. They also only measure the resistivity at or close to the surface of brickwork. For these reasons, resistivity meters are often not the most effective method of ascertaining the level of moisture within bricks and where this has led to a diagnosis of high moisture levels within brickwork, results should be interpreted carefully.

- *Capacitance meter:* The use of a capacitance meter is also a common investigative technique for surveying moisture within bricks. While capacitance meters can read moisture levels up to a depth of around 40 mm and are often more compatible with masonry than resistivity meters, their results can be disrupted by the presence of conductive materials under the surface of brickwork. Again, care should be taken in interpreting such results.

- *Gravimetric analysis:* A third method of identifying the moisture content of brickwork is to use the process of gravimetric moisture analysis. It should be noted that this is a destructive test and requires the removal of a sample of brick or mortar from a building. If brickwork is of particular heritage significance, such analysis may be considered inappropriate. When a sample of material has been removed from a wall, the weight of this material is recorded and then dried and weighed again. This provides information on the moisture content of the samples, and therefore of the masonry at different points.

- *Microwave moisture meters:* A relatively new technology, microwave moisture meters are increasingly being used to ascertain moisture levels within brickwork. Microwave radiation can penetrate up to 800 mm into brickwork and is largely unaffected by salt content within bricks and mortar. Microwave meters are non-destructive and are therefore more appropriate for sensitive heritage brickwork than destructive methods of analysis.

- *Thermal imaging:* Thermal imagining is a technology that has come into relatively common use in the surveying of traditional buildings. Thermal imagining cameras can be used to detect moisture within brickwork, as wet areas will be relatively lower in temperature compared with drier parts of a building. The use of thermography is discussed more fully in section 7.10.

Other methods of measuring moisture in brickwork are available, but are beyond the scope of this work. Orr (2020) provides an up to date summary of these. All methods of measuring moisture levels in brickwork may be regarded as having their own strengths and weaknesses. As with any form of specialist analysis, a clear understanding of the purpose of any testing is required before it is embarked upon. This will allow for the correct method to be selected and for the results to be interpreted correctly.

Fig. 7.5 The use of microwave moisture measurement – although in this case on a stone structure – can be an effective way to map moisture levels in brickwork (copyright HES).

7.4 Analysis of salts

As discussed in section 6.1, salts are both a cause and symptom of defects in traditional brickwork. The source of salts within brickwork is an important consideration when surveying brickwork affected by efflorescence. One of the most common sources of salts is cement mortar that has been used to repoint bricks which are highly porous, allowing the salts to travel from the mortar into the brick. Salts can also enter brickwork where chimneys and flues are present, a result of many years of burning fuel. These are often found concentrated in the brickwork around flues, but can also migrate to other parts of adjacent brickwork.

Salts can attract moisture from the air within a building, causing dampness. Some salts are highly hygroscopic and draw moisture into walls. The internal environment of many buildings caries a high level of humidity due to the actions of habitation, such as bathing and drying clothes. If salts are present within brickwork, these can draw moisture into the brickwork, making it appear much damper on the surface than may actually be the case. As discussed in section 7.4, the presence of salts can affect the proper interpretation of some methods of moisture analysis.

There are a number of reasons why salts may be tested as part of survey and assessment of brickwork. The testing of salts can be a useful investigative technique to aid the development of repair strategies and interventions such as poulticing to rectify high levels of salt within masonry. Many salts have different hydration phases and in this situation, where salts are identified (i.e. sodium sulphates), a non-aqueous conservation strategy might be applied to avoid any salt crystallisation damage from occurring within brickwork. Knowing the types of salts present within brickwork is important to help inform the most appropriate environmental conditions to avoid the salts dissolving and re-crystallising, as it is this action that can lead to deterioration of fired clay bricks.

The process of analysing salts may be summarised by the following steps: sampling, analysis and interpretation. Various questions can be answered by the analysis of salt affecting brickwork; the most significant in terms of survey and assessment is often to identify the types of salts found and potentially areas of brickwork that are worst affected, as this can provide information on their source.

In order to obtain samples of salt for analysis, two methods are principally used: removing a sample of the material directly from the surface of the brickwork; or using a poultice to draw salts from the brickwork to allow for analysis. Direct sampling would fall into the category of destructive testing, so there may be occasions where the poultice method is preferred if brickwork is of particular heritage significance. This method also allows the testing to be repeated following other interventions – for example, after the removal of cement render or pointing, which may have been the principal source of the salts.

A number of analytical techniques can be employed in the analysis of salts in brickwork. X-ray Diffraction can be used in the analysis of salts, although this requires careful calibration and will only detect ions that have crystallised to form minerals. Numerous techniques use water extraction procedures to obtain salts from masonry for analysis; it is beyond the scope of this book to cover them in detail, but Blaeuer 2005 discusses these in detail. The use of poultices to extract salts for analysis can utilise materials such as cellulose fibres or clay minerals. The extracted solution is then analysed for pH electrical conductivity and the content of a wide range of soluble salts.

The interpretation of the results of salt analysis can be complex, and requires clear definition of what constitutes high or low levels of salts within brickwork. The implications of this are, however, determined by the pore structure of the bricks: if bricks are highly porous, they may be able to accommodate the crystallisation of salts better than low-porosity bricks, which may be more susceptible to damage. The analysis of salts affecting brickwork is a complex process and, as with all specialist investigative techniques, it is important to define clearly at the outset what the testing is aiming to achieve. This type of testing is not required as a matter of course, but should be used where there is a clear need to ascertain the salts present within brickwork and to aid in identifying their source.

Fig. 7.6 Analysis of salt can be a useful tool for diagnosing the cause of defects in traditional brickwork.

7.5 Laboratory analysis of compressive strength and moisture absorption

It can be difficult to assess the compressive strength of brick masonry walls in
situ. Non-destructive testing generally does not give particularly reliable results.
Destructive testing, on the other hand, is often unachievable due to conservation considerations. Removing material from a wall to ascertain its strength
is, by its nature, a destructive process. Unless it is absolutely necessary, then,
achieving a precise compressive strength of a masonry wall should generally be
reconsidered. Traditional brickwork was built with a large margin of safety for
strength and brickwork performs well in compression; therefore, assuming it has
stood the test of time, it is likely that the strength of the masonry will be sufficient for the purpose to which is it being put. When taking samples of bricks
for this or other testing, it is vital that these are clearly recorded and labelled
regarding the location in a building from which they were taken. Whenever
material is removed from a building, it should always be remembered that this
may require statutory consent and could impact archaeological or heritage
significance.

Where historic bricks are being tested for compressive strength, the most common method used sees them placed in a compressive strength machine and pressure applied until the brick fails. Often five bricks are tested and an average taken to give a compressive strength rating for the bricks expressed in terms of Newtons per square millimetre (N/mm^2). It should be noted that this is a destructive test and will result in the loss of the bricks removed from a structure to allow the testing to be carried out. Care should also be taken to ensure that the bricks which are being tested are as representative of the overall brickwork as possible. If compressive strength is being tested, a detailed plan will need to be created to ensure that enough bricks are tested to give an accurate result for the brickwork as a whole. Given that brickwork generally performs well in compression, the removal of bricks for compressive strength testing offsite is likely to be a relatively rare part of the assessment of traditional brickwork.

The rate at which a brick absorbs water is a significant technical characteristic. This will determine how much moisture the brick is likely to take into its fabric during periods of wet weather externally, or high moisture loading and condensation internally. As moisture is a significant factor in decay mechanisms such as frost damage and salt mobilisation, this can have a significant impact on brickwork. It is also a factor in the sourcing of suitable replacement bricks, although that is beyond the scope of this work. Ascertaining the water absorption of traditional bricks is a further specialist laboratory analysis that can be undertaken to inform survey and assessment if required. As with compressive strength, however, it is not something that would be undertaken as a matter of course, but only where such information was of significance – most likely where future repair work was planned.

The method used to ascertain the water absorption of samples of traditional brickwork involves determining the weight change in bricks that have been dried to a constant weight at 105°C to 115°C, after their immersion in water. The bricks are initially washed to remove loosely adhering soiling with, where present, the remnants of any adhering bedding/jointing mortars also removed and the samples air dried. All bricks are then measured and photographed prior to being dried to a constant weight in an air circulating oven at 110°C, with the weight checked after 24 hours, 36 hours and 48 hours, until a constant weight is reached, in all samples, after 60 hours of drying.

After drying, the specimens are weighed to an accuracy of 0.1g (M1). The cooled and weighed samples are then totally immersed in water, at 27°C to 29°C for a period of 24 hours. After the soaking period, the samples are removed from the water, with excess water removed by wiping each sample with a damp absorbent cloth prior to the samples again being weighed to the same accuracy (M2). The water absorption is then calculated using the following formula: $W = M2 - M1 \times 100$ where W is the water absorption of the brick, as a percentage of its dry weight. This methodology has given consistent results of water absorption of traditional brick samples subjected to testing.

Table 7.2 Water absorption rates of traditional bricks

Brick type	Likely water absorption rate
Common	7–9% (lower absorption class)
	12.5–14.5% (higher absorption class)
Handmade	13–17%
Pressed facing	6–9%
Extruded	8–9% (lower absorption class)
	11–18% Higher absorption class)
Glazed	0.5% (fully glazed)
	7–11% (partially glazed)

Fig. 7.7 Extruded bricks such as these have been found in testing to have a generally high water absorption rate.

The water absorption results in Table 7.2 demonstrate the range that may be present in the water absorption of traditional bricks. The type of brick can clearly be seen to have an impact on this, although there are significant variations even within bricks of the same type. The testing of bricks for water absorption can be an important part of survey and assessment, but it is not something that should be undertaken lightly, as it is an invasive form of investigation.

7.6 Other laboratory analytical techniques

The most common materials taken for analysis from traditional brickwork are bricks themselves and samples of mortar. Other materials that may be removed for analysis include surface deposits, renders and paint samples. Where any material is being removed from a traditional brick building, it is important to consider the heritage significance and potential archaeological impact of any material. Any material that is removed from a traditional brick building should be clearly labelled regarding where in the building it came from. It should also be stored in such a way that contamination will not occur prior to any analysis or testing taking place. Detailed notes should be made on site to indicate where the sample has come from and what is hoped to be gained through the analysis. Permission from the building owner should always be secured prior to the removal of any material and in some cases statutory consent may be required prior to the removal of historic bricks, mortar or other materials.

It is possible to use a variety of laboratory-based material analysis techniques to ascertain information related to traditional bricks. X-ray diffraction is a laboratory-based technique that has already been discussed in relation to analysis of mortar and salts. It can also be used to analyse bricks themselves providing information related to mineralogy of the bricks and the clay from which they were made. This can be useful in establishing the provenance of bricks used in structures and in understanding decay mechanisms. Scanning electron microscopy (SEM) can also be used to analyse traditional bricks. This can reveal information such as the extent of vitrification and the firing temperature to which the bricks were subjected. Information related to porosity and pore structure of bricks can also be gained through SEM.

Thin sections of brick can also be taken in order to use petrographic microscopy to ascertain material properties of brick. Although this is a technique more commonly applied to analysis of stone and the matching of stone for use in repair work, much useful information can be gained by carrying out this type of analysis on samples of brick. Petrographic analysis can, for example, help in identifying the pore structure of bricks and the presence of voids and inclusions. The depth of vitrification, if present, can also be ascertained in this way. Inclusions such as those of lime or unfired material such as quartz, feldspar and sandstone fragments can also be discerned through petrographic analysis.

These analytical techniques are not something that would be undertaken as a matter of course during survey and assessment work. They can, however, yield a great deal of important information if required. The aim of any testing should always be clear – for example, to better understand the decay of bricks or to aid in repairs.

7.7 Monitoring of structural movement and cracks

A number of techniques can be employed when monitoring cracks in traditional brickwork. Simple dated pencil marks can be used to define the end point of a crack and can be drawn across a crack to detect differential movement across a crack. A simple crack width gauge can be used to ascertain the width and severity of a crack in masonry at a given time. A device commonly referred to as a "tell-tale" is one of the most common forms of long-term monitoring of cracks. Early types were made so as to break in half if a crack was live. A more advanced version of the traditional tell-tale is the Vernier crack marker, which measures changes in exterior cracks across a single plane over time. These are formed of two thin plastic plates attached on either side of a crack, one sitting on the other. If the crack is live, this will become apparent should the two plates move apart with the distance measured being indicative of the rate of expansion. Other types commonly used to monitor cracks in brickwork include the Avongard Tell Tale and Moire Tell Tales. The most common application of this is in relation to vertical cracks. However, there are forms of crack monitoring that can be used in other situations, including cracks found at the corner of brickwork. These should be fixed to brickwork itself rather than as an applied finish, which may detach.

More specialist techniques for monitoring cracks include the use of mechanical strain gauges, which are a further method of monitoring cracking in brickwork and take the form of metal discs or studs that can be fixed to either side of a crack. These allow measurements to be taken using a fixed or demountable strain gauge. An example of this is the DEMEC gauge. Where structural defects and cracks are found during visual inspection of brickwork, the use of monitoring of this type can be a helpful specialist technique in ascertaining the severity and current state of the crack.

Various systems exist that allow the use of sensors and data loggers to record structural movement and other forms of behaviour in traditional brickwork digitally. Off-site monitoring stations can also be used for this purpose. A variety of different material properties can be remotely monitored – for example, the moisture content of brickwork, either on the surface of the bricks or interstitially within the brickwork itself, using remote monitoring. The cost of this type of work, however, is likely to restrict it to all but the most sensitive or critical applications. It is essential that records are kept and reviewed regularly where such monitoring work is taking place and that the occupier manager of the building is aware of the work and does not remove or turn off the sensors.

Where it is noted that brick masonry has come to be out of line vertically ("out of plumb"), several methods can be used to confirm whether this is the case. The simplest method is the traditional plumb bob. This can be used to gain a quick idea of whether brickwork has become misaligned. Where greater accuracy is required, digital methods such as a device known as an optical plummet may be used to ascertain whether brickwork is misaligned. As with any specialist investigative technique, experience in correct interpretation of the results and the correct use and calibration of equipment is vital if the accuracy of results is to be relied upon. A number of other investigative techniques that can be helpful in assessing cracks are set out in section 7.7. The digital documentation techniques discussed in section 7.11 may also be helpful when investigating cracks and structural defects. There are also clearly benefits, where a BIM model is created, to incorporating the results of any crack monitoring that takes place.

The specialist investigative techniques discussed in this section relate to the monitoring and investigation solely of cracks and structural defects within brickwork. Clearly, where there are serious structural issues with any traditionally constructed building, investigation of other elements of the building's fabric will be required – particularly floors and roofs. It is beyond the scope of this work to consider the investigation of the building as a whole: for further information related to this, see the References and Further Reading section at the end of this chapter, particularly Beckman and Bowles (2004), Robson (1999) and Watt (2011). As discussed in section 1.5, it is always advisable to secure the services of a professional experienced in assessing traditionally constructed buildings.

Fig. 7.8 In this internal brick wall, moisture levels on the surface of the brickwork and 50 mm into the brick itself are being monitored. The moisture monitoring equipment links to a data logger. This work was carried out in relation to a retrofit project.

Fig. 7.9 The crack in this image runs through at least 10 courses of brickwork. It will be seen that it is cracking both mortar and bricks, a sign of a more serious problem than where hairline cracks affect mortar joints only. It is likely that a period of specialist monitoring will be required for a crack of this extent.

7.8 Other site-based investigative techniques

In addition to the techniques described above for measuring cracks, a wide range of other site-based investigative techniques are available for assessing traditional brickwork. The following list is not exhaustive, but gives a good idea of some of the specialist techniques that can be employed to traditional brickwork.

7.8.1 Free electromagnetic radiation

Free electromagnetic radiation techniques can be used to detect features hidden within brickwork. These can include service routes and flues built into brickwork. The technique can also be used to help identify cavity walls that may not be apparent from external inspection. This may be the case where brickwork has been rendered or where a one brick thick outer leaf has been used.

7.8.2 Microwave investigation

This investigative technique uses microwave energy transmitted into a material. The strength and direction of the reflected signal can then be used to determine the features of the structure. Where this is employed in the context of traditional brickwork, it can be used to identify voids in construction, changes in material and the presence of inbuilt reinforcement such as bond timbers. This should not be confused with the use of microwave investigation to measure moisture within walls, as it is a separate investigative technique.

7.8.3 Ultrasonic pulse velocity measurements

Similar to microwave investigation, ultrasonic pulse velocity measurement uses high-frequency sound waves to investigate traditional structures. Again, this can be helpful for locating voids, cracks or cavities, especially where the structure is of considerable mass.

7.8.4 Radiography

The use of x-rays or gamma rays is a further specialist method to look behind a façade or to see inside traditional brickwork. Again, this can be used to detect cracks, voids or hidden materials behind what is apparent on the surface. There are specific health and safety concerns with the use of this technique, however, and these should be borne in mind if this method is to be employed.

7.8.5 Fibre-optic investigation

Fibre-optic probes can be inserted into cavities or voids within traditional brickwork. The probes can take the form of endoscopes or borescopes, which deliver illumination to the subject and carry an image back to an eyepiece. Both endoscopes and borescopes can be used to inspect and record details within brickwork. They can be inserted either through existing cracks or means of access, such as vents, or by drilling small holes – preferably through mortar joints to avoid damage to brickwork. In instances where features within brickwork, such as flues, are being investigated, CCTV is a third option for specialist survey and recording.

7.8.6 Impulse radar

When investigating the condition of brickwork, a further specialised investigative technique is the use of impulse radar, also known as ground penetrating radar. Impulse radar is a non-destructive technique that utilises pulsed radio energy transmitted from an antenna held against the surface of a wall, which is then reflected back. As different construction materials absorb and reflect in different ways, impulse radar can help to locate voids, timber or metal within brickwork. This is a specialised investigative technique, however, which is not likely to be deployed in most surveys. It can be a useful technique where defects are seen within brickwork and the cause is not immediately apparent. For example, where cracks are thought to be caused by embedded metal corroding or timber decaying, impulse radar may be an appropriate specialist technique to implement.

7.8.7 Specialist survey for ferrous metal

The use of metal detection can aid in the diagnosis of defects in traditional brickwork where these are suspected to be a result of inbuilt or hidden metal construction materials. If metal detection is being undertaken, it is important that the equipment used is suitable for the application to which it is being put and is field calibrated to be effective in detecting wall ties or other in built metal. Electrical conductivity is the most common method used to detect metal within brickwork. This method can be used in combination with impulse radar investigation. Metal detection uses one of several electrical conductivity methods to detect the induced magnetic field of conductive targets buried within non-magnetic materials. In the context of traditionally constructed brickwork, this is likely to take the form of hoop iron reinforcement or, in early cavity brickwork, early metal cavity ties. Impulse radar may also be used to detect decayed cavity wall ties and other metal that has been built into brickwork. Where it is suspected that this is the cause of deterioration or decay within brickwork, the use of metal detection may be an appropriate investigative technique. As with any specialist survey, interpreting the results of metal detection is a skilled operation with a number of factors contributing to inaccurate results. For this reason, the use of experienced operatives is advisable.

Fig. 7.10 Various specialist investigative techniques can be used to locate corroding ferrous metal in traditional brickwork such as this cavity wall tie; a borescope was also used in this case.

7.9 Aerial survey

As discussed in Chapter 1, accessing traditional brickwork to gain the required information during survey work can be very challenging. The use of aerial survey methods, such as unmanned aerial vehicles (UAVs), is one way to access hard-to-reach brickwork. A number of regulatory requirements apply to the use of UAVs, especially in urban environments. For this reason, where such a survey technique is being used, the use of a fully qualified company to carry out this work is advisable. All relevant consents and legal requirements should be complied with.

Having gained the relevant information from the aerial survey, it needs to be interpreted. It is at this stage that the limitations of this survey method may become apparent. It is clearly difficult from an aerial survey to fully assess the condition of brickwork – for example, assessing whether mortar is friable.

Fig. 7.11 Aerial survey can be used to aid the survey of brickwork that is hard to access, such as at the Butt of Lewis Lighthouse, seen here. The Northern Lighthouse Board utilises UAV technology during survey of this structure (copyright Northern Lighthouse Board).

Where such limitations emerge and defects are suspected to be present, further access may be required to the brickwork. Still, the use of aerial survey can be an important specialist investigative technique when surveying hard-to-access traditionally constructed brick buildings.

7.10 Thermography

The use of thermal imaging or infrared thermography is a further specialist and non-destructive technique for investigating defects within brickwork. Thermal cameras work by detecting infrared radiation related to the temperature of an object or part of a structure. A thermal image is one where each pixel of the image records a temperature, giving a detailed impression of the temperature of the structure at various points. The thermal imaging camera measures infrared radiation from a surface and converts this into a temperature utilising the material property emissivity. The detailed mechanics of thermal imaging are discussed more fully in Young (2015).

Thermography can be useful in a number of ways in the survey and assessment of brickwork. The technique can help to detect voids within brickwork and

Fig 7.12 The use of thermography to analyse the condition of a building can be seen here. Thermography has been used to ascertain the extent of dampness caused by a leaking downpipe (copyright HES).

Fig. 7.13 The thermographic image has been overlaid onto a photograph showing 20th century brickwork. This can be an effective way of expressing the results of thermography (copyright HES).

can be particularly useful in this capacity where masonry may be of composite construction – for example, a brick-lined rubble wall. As discussed in section 6.7, where brickwork is not well bonded with rubble or ashlar backing or even where facing bricks are backed with bricks of another type, instability can lead to hidden voids. This would be more difficult than using thermography to reveal render that was coming loose from a wall, as it will require the temperature within the hidden void to rise and then the wall to be thermally imaged as it cools to identify voids where the heat remains trapped. This may require the use of a specialist thermography company and the use of heaters in some instances.

Thermography can also be used to detect the presence of moisture and dampness within brickwork, as damper areas will have a lower temperature than those that are dry. Care is required in the use of infrared thermography, however, to ensure that data is gathered correctly, and that this is interpreted in the right way. The use of infrared thermography is certainly of potential value when investigating defects in brickwork, although it would be regarded as a specialised intervention.

7.11 Digital documentation and photogrammetry

Advances in the technology of applying digital documentation and recording to standing structures have led to the use of a variety of these being used in the survey, assessment and recording of traditional brickwork. Despite being considered a specialist technique in the context of this document, laser scanning is becoming increasingly mainstream with regard to traditional brick buildings.

One of the most common laser scanning techniques applied to traditional brickwork is terrestrial laser scanning (TLS). TLS utilises a technique known as LiDAR (light detecting and ranging) to acquire the measurements of a structure. This essentially works by firing a beam of light from a laser at a surface and measuring its return properties to gauge the distance between its origin and the object surface. Scanners spin a mirror to take measurements along vertical lines of data as they rotate, capturing the surrounding environment with a near-spherical line of sight. However, for some buildings there may be insufficient space or access to use this technique, in which case technologies such as mobile scanning or unmanned aerial vehicles (UAVs) may be employed to gather the relevant data. The level of detail required will inform the technique chosen.

Laser scanning can be used to create what is termed a "point cloud" of a structure's surface. This is a dense array of 3D coordinates gathered from a structure. It can then be used for a number of applications. At the most basic level, dimensions can be taken and two-dimensional architectural drawings can also be produced – for example, of a brick arch that may require repair. Areas and volumes can also be ascertained in this way. The same dataset can also, in some instances, potentially be used to create a 3D replica of a building component to aid in repair work. Although this is less likely to be applicable to brickwork than traditional building materials such as cast iron, it could potentially be useful where carved bricks or complex specials are being surveyed ahead of replacement. There is also potential in this regard where terracotta or faience elements are present. A sufficiently high-resolution dataset might enable visualisation of a masonry surface or building element, revealing structure and details that were not previously visible.

Where part of a structure is repeatedly scanned over time, there is potential to use the output of this to monitor decay or building failure – for example, cracks within brickwork or, if sufficiently detailed, the spread of spalling across an area of brickwork. The point cloud can also be used to overlay other monitoring data on such as photography or thermography. As a source of digital information, there are clearly a number of ways in which the output of laser scanning can link to BIM, as discussed in section 1.4.2. Laser scanning is commonly used as the underlying dimensional dataset for the BIM process for existing traditional and modern buildings.

Laser scanning is often used in conjunction with photogrammetry. In its broadest sense, photogrammetry is the technique of using photography to derive three-dimensional information based on the differences between perspectives of building elements across two or more photographic images. Increasingly lower cost options such as structure from motion photogrammetry make the use of digital methods of capturing information related to traditional brick-work more viable. Structure from motion photogrammetry can be produced using only a camera and appropriate software to process the data, although a good understanding of what the process entails is also important. The use of high-resolution images results in detailed image "textures" in addition to the 3D surface information, which in some cases can more clearly show surface details and conditions. When using this form of photogrammetry, all access-ible and visible surfaces require to be captured from as many unique angles as possible. This will ideally involve taking photographs from near ground level, from standing height and from an elevated position as well as from different angles. Photogrammetry can be a useful technique for recording and presenting information related to traditional brick structures and can link well with laser scanning data and BIM in many instances.

This short section has provided only a brief introduction to the use of laser scanning and other related technology with regard to traditional brickwork. Several resources can be referred to when considering applications of this form of technology to traditional brickwork, most notably Frost (2018) and Boardman (2018). Laser scanning and other digital documentation techniques have a number of potential uses in the survey and recording of traditional brickwork. From detailed recording of significant heritage assets to the use of more basic technology to explore and record the condition of buildings, these developments are likely to be a significant part of the survey and recording of traditional brickwork in the future.

7.12 Case study 8: Central Presbyterian Church, Chambersburg, Pennsylvania, United States – analysis of surface coating

With thanks to Bill Revie, CMC Stirling for contributing text and images.

The building considered in this case study is the Central Presbyterian Church, Chambersburg, Pennsylvania in the United States. The church was constructed in 1868, with an archive image of 1870 indicating that the building was painted white on at least three sides. A severe fire in the 1830s caused significant damage, particularly to the south side of the building, which archive images suggest was not painted white. Material samples were sent to Bill Revie, who kindly submitted the material presented in this case study. One of the principal aims of the testing was to establish the composition of a colour wash on the surface of the brick.

For the scientific analysis, a sample of a single, whole brick was sent to the laboratory. This was examined using a stereo-binocular microscope at magnifications up to x40. It was also exposed to a series of droplet tests using various reagents and indictor solutions. Petrographic thin sections were also prepared, including one that showed the outer "painted" surface of the brick.

The brick has the appearance of a hand-moulded brick, where the brick was hand-pressed into a mould and levelled with a hand tool. The abundance of quartz grains on the surfaces of the brick would suggest that the mould had been lightly sanded before the clay was cast in, or the unfired brick had been lightly sand rubbed before firing. To establish whether these features could assist in establishing a date of manufacture for the brick, it would be necessary to have knowledge of the local practices followed in brick manufacture at the time of the original construction – that is, whether sand moulding was practised during the period in which the brick was thought to have been manufactured.

Surface scrapings were taken from each of the outer, inner and bed faces of the brick for analysis by x-ray powder diffraction (XRD). In preparation for the analysis, a scraping was made from the surface, with the collected dust ground in an agate mortar and pestle. The powder was then deposited as a thin film on a glass slide in the form of an acetone slurry, which was evaporated to dryness in readiness for presentation in the XRD. Further samples were prepared, following the same procedure, from a scraping taken from the rear soot-stained surface, and from a mortar-stained area on the top bed face.

The basis of the examination of samples by XRD is that when a crystalline material is scanned by a beam of x-rays under controlled conditions, it produces a diffraction pattern that is characteristic and unique to the material under examination. In a mixture of crystalline materials, each component generates a distinct diffraction response allowing identification of the materials present. As the laboratory analysis programme progressed a sub-sample, cored from the surface of the brick, was also obtained and analysed by x-ray fluorescence spectrometry to identify the chemical composition of the surface coating and to establish whether this changed with depth.

The front face of the brick was found to have retained a surface coating that was strongly bonded/fused to the surface of the brick. This coating was found to be virtually impermeable, with water droplets placed on its surface maintaining a stable meniscus for a considerable period of time, whereas water droplets placed on an uncoated area of brick were absorbed and diffused across the surface, within five to ten minutes. Although the coating displayed a crazed finish, there was no evidence to suggest preferential absorption along the map crack boundaries. Locally, the surface coating had been disrupted with loss of small patches, which exposed the underlying brick. At these locations, very localised light fabric disruption was observed, with a loss of brick surface materials.

On the basis of ad hoc droplet tests, the coating was found not to be dissolved or disrupted in the presence of dilute acid solutions, alkali, organic solvents or oxidising agents. The coating was therefore indicated to be relatively inert, which may infer that it had been exposed to the atmosphere and weathering for a long time, resulting in the complete oxidisation of any organic components, along with the alteration/conversion of any lime-based materials, such as a limewash or other traditional wall coating.

Examination of the thin sections again confirmed that the coating had penetrated into the surface of the brick, but in general it appeared to have formed a crust on the outer surface of the brick. The coating, when examined in the polarised light microscope, was noted to be mainly crystalline; however, a sharp delineation between the coating and the brick surface was not clear. However, there were indications that some of the coating materials had been absorbed into the surface of the brick, in porous areas, with additional penetration into large-surface connected pores and along micro-crack pathways.

The XRD analysis indicated that the dark surface coating on the front of the brick is composed mainly of lead-based compounds, with lead sulphate dominating, trace proportions of lead sulphide (PbS) and elemental lead (Pb). This, and the presence of lead carbonate, would suggest that the original coating had been a lead-based paint – probably white lead – with subsequent sulphate reaction resulting in the darkened surface now apparent. The source of the sulphate could be the brick itself or environmental pollution, but on the basis of this examination and analysis, it is most likely the latter. An analysis by x-ray fluorescence spectrometry (XRF) confirmed that the coating was composed mainly of lead, sulphur and calcium, with other minor components. The appearance of the coating shows that it was applied as a paint onto the brick/wall surface and not as a "glaze" on individual bricks.

This short case study, part of a larger body of testing on bricks and mortar from the church, demonstrates a practical application of laboratory-based material testing on traditional bricks. It will be seen that such testing can provide invaluable information regarding surface coatings on brick and has a wide range of potential applications when seeking to survey and assess traditional brick structures.

Fig 7.14 Close-up of a water droplet test on an area of the "painted" surface, sometime after application (copyright Bill Revie CMC).

7.13 Case study 9: As-built BIM for 15th century Chinese brick structure

With thanks to Sun Zheng and Jiangtao Xie for permission to reproduce material from the paper Sun, Z., Xie, J., Zhang, Y. and Cao, Y. (2019), "As-Built BIM for a Fifteenth-Century Chinese Brick Structure at Various LoDs", ISPRS International Journal of Geo-Information, *vol. 8, no. 12, p. 577.*

The building that forms the basis of this case study is the Stele Tower, Nanjing, China, which was built during the Ming Dynasty (1368–1644 CE). This period witnessed considerable development in the use of brick for construction in China. The Great Wall of China was extensively strengthened and rebuilt for the third time in history, and most of the structures observed today in this context originated during this period. As the capital of the Ming Empire before 1421 CE, Nanjing retains many brick-built ruins, including royal tombs, temples, civil infrastructure and city walls. The structure discussed in this case study dates to the 15th century and was surveyed using a range of innovative techniques as part of a research project to assess the effectiveness of presenting survey results in an "as-built BIM" format.

Building information modelling (BIM) is a component-based modelling environment that is semantically structured. As-built BIM refers to a BIM representation of the "as is" conditions of built heritage at the time of a survey. Although developed with a range of construction and engineering applications in mind, BIM is increasingly finding applications related to historic and traditional buildings, something often referred to as historic BIM (HBIM). It is considered a potential platform to bridge the gap between advanced measurement technology and various conservation-oriented applications. As-built BIM can be used to record and depict damage, missing elements of a building's fabric and other decay using advanced measurement technologies. BIM is discussed in the context of brickwork in section 1.4.2 of this book.

One of the crucial factors when looking at how to use BIM to present survey results for traditional brick structures is the "level of development" that is to be used. In relation to the building under consideration in this case study, three levels of development were presented:

- a host model linked with raster images composed using orthoimage and relief maps (LoD 1)
- an as-built volume with semantic skins (LoD 2)
- a brick-by-brick model with custom industry foundation class parameters at local areas (LoD 3).

In the surveying of the brick structure, a range of digital documentation techniques were utilised. These included terrestrial laser scanning (TLS) and photogrammetry, a combination of these methods being used to achieve high-resolution (HR) documentation with high geometrical accuracy (mm level) and chromatic fidelity (24-bit colour space). Where surveyed spaces were too narrow to use TLS and photogrammetry, technologies such as mobile scanning and spherical cameras were used.

Graphically representing measurements obtained using such advanced methods or recording can be more challenging than traditional methods of measuring and surveying. In this case study, both terrestrial laser scanning and photogrammetry were employed when measuring the Stele Tower. In addition to accuracy, the portability and efficiency of the recording methodology were considered when conducting measurements as tourist footfall was an important consideration.

For the recording work, a Leica BLK 360 Imaging Laser Scanner was used to scan the entire building and generate a global mesh model from 21 separate scans. The mean error of registration (based on natural features) was less than 3 mm. The segments of textured mesh surfaces were generated using photogrammetry. Images were obtained using hand-held cameras with convergent camera networks and automatically aligned using Agisoft PhotoScan. The generated mesh surfaces were approximately registered using the control points extracted from the terrestrial laser, which were scanned before being imported into the BIM environment. Damage types were recorded consisting of two values ("crack" and "impact damage"), based on preliminary observations by the naked eye.

The three levels of development produced from the work above to be input into BIM ranged significantly in detail. Even at the most detailed level, however, it was noted that "the obtained model was not strictly as built, as the brick units were generated based on the assumption of stretcher bond without addressing their irregular arrangement caused by slight change in brick length. Manual adjustments would require high labour intensity." This is certainly a potential issue for the use of BIM regarding the presentation of digital survey work relating to traditional brickwork that is formed of many relatively small masonry units.

The direct application of what was noted as LoD 1 is in situ maintenance through the combined use of photogrammetry and BIM on portable devices. This is a promising utilisation given Nanjing's 35 km-long city wall, built during the Ming Dynasty. Currently, the damage documentation in daily maintenance relies mostly on photographs. The potential for the use of terrestrial laser scanning to detect damage and deterioration over time is certainly something that could prove beneficial in the future. BIM models at LoD 1 allow for web-based damage annotation and data sharing for these brick-built heritage sites.

The principal question that the work at Stele Tower sought to answer was whether a brick-by-brick model with a consequently large data file size was required for those surveying and maintaining a traditional brick structure. The answer, as with so much else regarding the surveying of traditional brickwork, is dependent on the building being assessed. If detailed information regarding the condition of brickwork is required to be presented – for example, if representing spalled bricks ahead of repair work – a method of presenting survey results that can present a brick-by-brick model is required. However, if the representation only requires showing broader areas that need further survey work, this level of detail would not be required.

If advanced methods of scanning traditional brick buildings are being used and the results are input into BIM, any restrictions on the level of detail should be highlighted, especially if this impacts the technical or aesthetic features of traditional brickwork as presented in Chapters 3 and 4. For example, if bond is not accurately represented or brick size is standardised, this should be noted clearly.

The results of this case study are revealing for those who may have cause to represent survey work in an as-built BIM format. In the simplest version, LoD 1, this could be used to present a web-based workflow for brick-damage annotations; as-built dimensions could be extracted from LoD 2; and LoD 3 enables attributes such as damage types to be attached at the level of detail of individual bricks. Clearly, the most detailed version is likely to prove the most durable in the long term, and will be most effective in giving detailed results in terms of issues such as the number of bricks that may need to be replaced or areas that need to be repointed. However, the time required to produce such a detailed model and the potential computer storage space required to host such a model are factors that need to be considered when assessing the level of detail regarding the use of BIM when presenting the results of survey work on traditionally constructed brickwork. Overall, this case study proves there is certainly considerable potential for BIM to be used in conjunction with laser scanning techniques, but this requires careful planning and management.

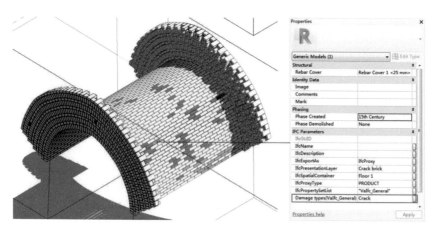

Fig. 7.15 An extract from the BIM model produced of the Stele Tower, Nanjing, China. The bricks were annotated to show damage by their damage types in BIM. "Crack" bricks and "impact damage" bricks were shown in red and green respectively (copyright Sun Zheng et al.).

References and further reading

Barham M. (2019), "Mortar Analysis", in *Building Conservation Directory*, Tisbury: Cathedral Communications.

Beckman P. and Bowles R. (2004), *Structural Aspects of Building Conservation*, London: Butterworth-Heinemann.

Bedford J. (2017), *Photogrammetric Applications for Cultural Heritage*, London: Historic England.

Blauer-Bohm C. (2005), "Quantitative Salt Analysis in Conservation of Buildings", *Restoration of Buildings and Monuments: An International Journal*, vol. 11, no. 6, pp. 1–10.

Boardman C. (2018), *3D Laser Scanning for Heritage*, London: Historic England.

Burkinshaw R. and Parrett M. (2003), *Diagnosing Damp*, London: RICS.

Freestone, I.C. (1995), "Ceramic Petrography", *American Journal of Archaeology* vol. 99, pp. 111–15.

Freestone I.C. and Middleton A.P. (1991), "Mineralogical Applications of the Analytical SEM in Archaeology", *Mineralogical Magazine* vol. 51, pp. 21–31.

Frost A. (2018), *Short Guide: Applied Digital Documentation in the Historic Environment*, Edinburgh: Historic Environment Scotland.

GB Geotechnical Limited (2001), *Non-destructive Investigation of Standing Structures*, Edinburgh: Historic Scotland.

Hall C. and Hoff W.D. (2012), *Water Transport in Brick, Stone and Concrete*, London: Spon Press.

Historic England (2016), *Measuring Moisture Content in Historic Building Materials*. London: Historic England.

Orr S. (2020), *Moisture Measurement in the Historic Environment*, Edinburgh: Historic Environment Scotland.

Pavia S. (2006) "The Determination of Brick Provenance and Technology Using Analytical Techniques from the Physical Sciences", *Archaeometry* vol. 48, no. 2, pp. 201–18.

Pearson C. (2011), *A BSRIA Guide: Thermal Imaging of the Building Fabric*, Bracknell: BSRIA.

Pender R., Ridout B. and Curteis T. (eds) (2014), *Practical Building Conservation: Building Environment*, Farnham: Ashgate.

RICS and Historic England (2018) *Investigation of Moisture and its Effects in Traditional Buildings*, London: RICS.

Robson P. (1999), *Structural Repair of Traditional Buildings*, Shaftesbury: Donhead.

Sun Z., Xie J., Zhang Y. and Cao Y., (2019), "As-Built BIM for a Fifteenth-Century Chinese Brick Structure at Various LoDs", *ISPRS International Journal of Geo-Information*, vol. 8 no. 12, p. 577.

Watt D. (2011), *Surveying Historic Buildings*, London: Routledge.

Young M. (2015), *Short Guide on Thermal Imaging in the Historic Environment*, Edinburgh: Historic Environment Scotland.

Bibliography

A number of works should be highlighted as being of particular overarching significance, in particular:

Hendry A., McCaig I., Willett C., Godfraind S. and Stewart J. (2015), *Historic England Practical Building Conservation: Earth, Brick and Terracotta*, Farnham: Ashgate.
Lynch G. (2007), *The History of Gauged Brickwork*, London: Elsevier.
Robson P. (1999), *Structural Repair of Traditional Buildings*, Shaftesbury: Donhead.
Watt D. (2011), *Surveying Historic Buildings*, London: Routledge.

The website of the Building Conservation Directory has many relevant articles, a number of which helped inform this work. See www.buildingconservation.com
The publication of the British Brick Society, *Information*, also has a considerable amount of material relevant to the history and development of brickwork.

Adams H. (1906), *Building Construction,* London: Cassell.
Ainslie (1842), "Ainslie Patent Brick and Tile Machine", *The Civil Engineer and Architects Journal, Volume 5*, London: Groombridge.
Allen G., Allen J., Elton N., Farey M., Holmes S., Livesey P. and Radonjic M. (2003), *Hydraulic Lime Mortar for Stone, Brick and Block Masonry*, Shaftesbury: Donhead.
Allen J. (1893), *Practical Building Construction*, London: Crosby Lockwood.
Ambrose J. and Tripeny P. (2006), *Simplified Engineering for Architects and Builders*, Oxford: Wiley.
Antonopoulou, S.B. (2017). *BIM for Heritage: Developing a Historic Building Information Model*. London: Historic England.
Arnot H. (1779), *The History of Edinburgh,* Edinburgh: Creech.
Ashpitel A. (1855), *Town Dwellings*, London: Weale.
Ashpitel A. (1867), *Treatise on Architecture*, Edinburgh: Adam and Charles.
Ashurst N. (1994), *Cleaning Historic Buildings: Volumes 1 and 2*, London: Donhead.
Ashurst N. (2008), *The Investigation, Repair and Conservation of the Doulton Fountain, Glasgow,* Edinburgh: Historic Scotland.
Ashurst N. (2011), "Cleaning Brickwork and Terracotta", *Building Conservation Directory*, Tisbury: Cathedral Communications.
Atkinson W. (1805), *Views of Picturesque Cottages with Plans,* London: Gardiner.
Austin J. (1862), *A Practical Treatise on the Preparation, Combination and Application of Calcareous and Hydraulic Limes and Cements*, London: Turner.
Ayres J. (1998), *Building the Georgian City*, New Haven, CT: Yale University Press.
Bakewell (1840) "Notice of Patent Brick Machine", *Architects and Engineers Journal*, London: Groombridge.

Barham M. (2019), "Mortar Analysis" in *Building Conservation Directory*, Tisbury: Cathedral Communications.

Beckman P. and Bowles R. (2004), *Structural Aspects of Building Conservation*, London: Butterworth-Heinemann.

Bedford J. (2017), *Photogrammetric Applications for Cultural Heritage*, London: Historic England

Bennett B. (2005), "The Development of Portland Cement", in *The Building Conservation Directory,* Tisbury: Cathedral Communications.

Bidwell T.G. (1977), *The Conservation of Brick Buildings*, Windsor: Brick Development Association.

Blauer-Bohm C. (2005), "Quantitative Salt Analysis in Conservation of Buildings" *Restoration of Buildings and Monuments* vol. 11, no. 6, pp. 1–10.

Boardman C. (2018), *3D Laser Scanning for Heritage*, London: Historic England.

Bonnell D., B. Butterworth (1950), *Clay Building Bricks of the United Kingdom*, London: HMSO.

Bonshor R. and Bonshor L. (1995), *Cracking in Building*, Bracknell: BRE.

Braby (1893), *Catalogue of Malleable and Cast Iron Goods Produced by F. Braby and Co. of Glasgow,* Glasgow: Braby and Co.

British Standards Institute (2001), *BS EN 459–1:2001, Building Lime: Definitions, specifications and conformity criteria*, London: British Standards Institute.

British Standards Institute (2002), *BS EN 13139:2002, Aggregates for Mortar*, London: British Standards Institute.

British Standards Institute (2013), *BS 7913:2013 Guide to the Conservation of Historic Buildings*, London: British Standards Institute.

Brocklebank I. (2012), *Building Limes in Conservation*, Shaftesbury: Donhead.

Brown A. (2017), "Hot Mixed Mortars", in *The Building Conservation Directory,* Tisbury: Cathedral Communications.

Brick Development Association (2000), *BDA Guide to Successful Brickwork*, London: Butterworth-Heinemann.

Brunskill R.W. (2009), *Brick and Clay Building in Britain*, New Haven, CT: Yale University Press.

Burkinshaw R. and Parrett M. (2003), *Diagnosing Damp*, London: RICS.

Burn R.S. (1871) *The New Guide to Masonry, Bricklaying and Plastering, Theoretical and Technical,* Glasgow: McCready, Thompson and Niven.

Campbell J.W.P. (2003), *Brick: A World History*, London: Gollancz.

Chapman S. and Fidler J (2000), *The English Heritage Directory of Building Sands and Aggregates*, Shaftesbury: Donhead.

Christy A. (1882) *A Practical Treatise on the Joints Made and Used By Builders*, London: Crosby Lockwood.

Clifton-Taylor A. (1987), *The Pattern of English Buildings*, London: Faber.

Copp S. (2009) "The Conservation of the Old Schoolhouse at Logie, Montrose", in *Vernacular Buildings 32*, Forfar: Scottish Vernacular Buildings Working Group.

Cox A. (1979) *Brickmaking: A History and Gazetteer Survey of Bedfordshire*, Bedford: Bedfordshire County Council.

Curtis R. and Hunnisett J. (2017). *Climate Change Adaptation for Traditional Buildings*, Edinburgh: Historic Environment Scotland.

Darbishire H. (1865), "The Introduction of Coloured Bricks in Elevations", *The Civil Engineer and Architects Journal,* vol. 28, pp. 66–9.

Davis C.T. (1889) *A Practical Treatise on the Manufacture of Bricks, Tiles, Terra-Cotta, etc.*, Philadelphia, PA: Henry Carey Baird.

Dearne T. (1821) *Hints on an Improved Method in Building*. London: Harding.

De Vekey R.C. (2000). *BRE Digest 329: Installing Wall Ties in Existing Construction*, Bracknell: BRE.

Dickinson P. and Thornton N. (2004), *Cracking and Movement in Buildings*, Coventry: RICS.

Dobson E. (1850), *A Rudimentary Treatise on the Manufacture of Bricks and Tiles*, London: John Weale.

Douet J. (1990), *Going Up in Smoke:The History of the Industrial Chimney*, London:Victorian Society.

Douglas G. (1985) *A Survey of Scottish Brickmarks*, Glasgow: SIAS

Douglas J. (1995), "Basic Diagnostic Chemical Tests for Building Surveyors", *Structural Survey*, vol. 13, no. 3, pp. 22–7.

Earl J. (2003), *Building Conservation Philosophy*, Shaftesbury: Donhead.

English Heritage (1994), "The History, Technology and Conservation of Architectural Ceramics", *Conference Papers, UKIC/English Heritage Symposium*, London: English Heritage.

Espinosa-Marzal R.M. and Scherer G.W. (2010), "Advances in Understanding Damage by Salt Crystallisation", *Accounts of Chemical Research*, vol. 43, no. 6, pp. 897–905.

Eyles V.A. and Anderson J.G.C. (1946), *Brick Clays of North-east Scotland, Part 1: Description of Occurrences*. Wartime pamphlet. London: Geological Survey and Museum.

Fairbairn W. (1854) *On the Application of Cast and Wrought Iron to Building Purposes*, London: Weale.

Firman R.J. 1994 "The Colour of Brick in Historic Brickwork", *British Brick Society Information*, vol. 61, pp. 3–9.

Firman R.J. and Firman P.E. 1967 "A Geological Approach to the Study of Medieval Bricks", *Mercian Geologist* vol. 2, no. 3, pp. 299–318.

Forsyth M. (ed.) (2007), *Structures and Construction in Historic Building Conservation*, Oxford: Blackwell.

Franke L. and Shuman I. (1998), *Damage Atlas: Classification and Analysis of Damage Patterns Found in Brick Masonry*, Stuttgart: Fraunhofer IRB Verlag.

Freestone I.C. (1995), "Ceramic Petrography", *American Journal of Archaeology*, vol. 99, pp. 111–15

Freestone I.C. and Middleton A.P. (1991), "Mineralogical Applications of the Analytical SEM in Archaeology", *Mineralogical Magazine*, vol. 51, pp. 21–31.

Frew C. (2007), "Pointing with Lime", *The Building Conservation Directory*, Tisbury: Cathedral Communications.

Frost A. (2018), *Short Guide: Applied Digital Documentation in the Historic Environment*, Edinburgh: Historic Environment Scotland.

Gailey A. (1984), *Rural Houses of the North of Ireland*, Edinburgh: John Donald.

GB Geotechnical Limited (2001), *Non-destructive Investigation of Standing Structures*, Edinburgh: Historic Scotland.

Gibbons P. (2003) *Preparation and Use of Lime Mortars*, Edinburgh: Historic Scotland.

Glover P. (2006), *Building Surveys*, Oxford: Elsevier.

Goudie A.S. and Viles H.A. (1997), *Salt Weathering Hazards*, Chichester: John Wiley.

Gourlay C. (1903) *Elementary Building Construction and Drawing*, Glasgow: Blackie.

Greenhalgh, R. (1947), *Modern Building Construction*, London: New Era

Grier W. (1852), *The Mechanics Calculator*, Glasgow: Blackie.

Hall C. and Hoff W.D. (2012), *Water Transport in Brick, Stone and Concrete*, London: Spon Press.

Hammond M. (1990), *Bricks and Brickmaking*, Aylesbury: Shire.

Haskoll W.D. (1857), *Railway Construction*, London: Achley.

Hasluck P. (1905), *Practical Brickwork*, London: Cassell.

Hasluck P. (1906), *Iron, Steel and Fire Proof Construction*, London: Cassell.

Hendry A., McCaig I., Willett C., Godfraind S. and Stewart J. (2015), *Historic England Practical Building Conservation: Earth, Brick and Terracotta*, Farnham: Ashgate.

Historic England (2011), *Practical Building Conservation: Mortars, Renders and Plasters*, Farnham: Ashgate.

Historic England (2016), Measuring Moisture Content in Historic Building Materials, Farnham: Ashgate.

Hume J. (1976), *The Industrial Archaeology of Scotland II*, London: Batsford.

Hunnisett J. and Torney C. (2013), *Lime Mortars in Traditional Buildings*, Edinburgh: Historic Environment Scotland.

ICS (1898) *Brickwork, Terracotta and Tiling*, London: Wyman.

Jenkins M. (ed.) (2009), *Building Scotland*, Edinburgh: Birlinn.

Jenkins M. (2014), *Scottish Traditional Brickwork*, Edinburgh: Historic Scotland.

Jenkins M. (2018), *The Scottish Brick Industry*, Catrine: Stenlake.

Jenkins M. and Curtis R. (2021), *Retrofit of Traditional Buildings*, Edinburgh: Historic Scotland.

Johnston S (1992), "Bonding Timbers in Old Brickwork", *Structural Survey*, vol. 10, no. 4, pp. 335–62.

Jones E. (1986), *Industrial Architecture in Britain 1750–1939*, London: Batsford.

Langley B. (1774), *The London Prices of Bricklayers' Materials and Works*, London: John Wren.

Larsen E.S., Neilson C.B. (1990), "Decay of Bricks Due to Salt", *Materials and Structures* vol. 23, no. 1, pp. 16–25.

Lloyd N. (1925), *A History of English Brickwork*, London: Institute of Clayworkers

Lynch G. (1994), *Brickwork: History, Technology and Practice*. London: Donhead.

Lynch G. (2006), *Gauged Brickwork: A Technical Handbook*, London: Routledge.

Lynch G. (2006), "The Colour Washing and Pencilling of Historic English Brickwork" *Journal of Architectural Conservation*, vol. 12, no. 2, pp. 63–80.

Lynch G. (2007), *The History of Gauged Brickwork*, London: Elsevier.

Lynch, G. (2007), "The Myth in the Mix: The 1:3 Ratio of Lime to Sand", in *The Building Conservation Directory*, Tisbury: Cathedral Communications.

Lynch G. (2012), "Tudor Brickwork", in *The Building Conservation Directory*, Tisbury: Cathedral Communications.

Macfarlane D. (1860), *A Description of the Colossal Chimney at Port Dundas with Remarks on the Construction of Buildings of a Similar Kind*, Glasgow: Russell.

McCaig, I. (ed.) (2013), *Practical Building Conservation: Conservation Basics*. Farnham: Ashgate.

McGregor C. (1996), *Earth Structures and Construction in Scotland*. Edinburgh: Historic Scotland.

Morton T. (2008), *Earth Masonry Design and Construction*, Bracknell: BRE.

Moxon J. (1703), *Mechanik Exercises; or, The Doctrine of Handy-works*, New York: Praeger.

Nicholson J. (1825), *The Operative Mechanic and British Machinist*, London: Sidney.

Nicholson P. (1834), *The Builder and Workman's New Directory*, London: Taylor.

Nicholson P. (1838), *Practical masonry, bricklaying and plastering*, London: Kelly.

Nicholson P. (1860), *The Guide to Railway Masonry Containing a Complete Treatise on the Oblique Arch*, London: Spon.

Oglethorpe M. (1993), *Brick, Tile and Fireclay Industries in Scotland*, Edinburgh: RCAHMS.

Orr S. (2020), *Moisture Measurement in the Historic Environment*, Edinburgh: Historic Environment Scotland.

Oxley R. (2003), *Survey and Repair of Traditional Buildings: A Sustainable Approach*, Shaftesbury: Donhead.

Partington C.F. (1825), *The Builder's Complete Guide*, London: Sherwood Stephen.

Pasley C.W. (1826), *Practical Architecture*, Chatham: Royal Engineering Establishment.

Pavia S. (2006), "The Determination of Brick Provenance and Technology Using Analytical Techniques from the Physical Sciences" *Archaeometry* vol. 48, no. 2, pp. 201–18.

Paxton R. (2007), *Scotland – Lowlands and Borders,* Birmingham: ICE.

Pearson C. (2011), *A BSRIA Guide: Thermal Imaging of the Building Fabric*, Bracknell: BSRIA.

Pender R., Ridout B. and Curteis T. (eds) (2014), *Practical Building Conservation: Building Environment*, Farnham: Ashgate.

Rawlinson R. (1859), *Designs for Factory, Furnace, and Other Tall Chimney Shafts,* London: Kell Brothers.

Robson P. (1999), *Structural Repair of Traditional Buildings*, Shaftesbury: Donhead.

Richardson B. (2000), *Defects and Deterioration in Buildings*, London: Taylor and Francis.

RICS (2018), *Surveying Safely*, London: RICS.

RICS and Historic England (2018), *Investigation of Moisture and Its Effects in Traditional Buildings,* London: RICS.

Rivington (1905), *Rivington's Notes on Building Construction Volume 1*, London: Longmans.

Salmon W. (1748), *The London and Country Builders Vade Mecum; or, the Compleat and Universal Architect's Assistant* , London: Hodges.

Sanderson K. (1990), *The Scottish Refractory Industry*, Edinburgh: Author.

Scott Burn R. (1870), *The New Guide to Masonry, Bricklaying and Plastering, Theoretical and Technical,* Glasgow: McCready, Thompson and Niven.

Searle A. (1913), *Cement, Concrete and Bricks*, London: Constable.

Shaw E. (1832), *Operative Masonry*, Boston, MA: Capen and Lyon.

Smeaton A.C. (1836), *The Builder's Pocket Manual*, London: Tudor.

Smeaton J. (1837), *Reports of the late John Smeaton, Vol. 1*, London: Michael Taylor.

Sowden A. (1990), *The Maintenance of Brick and Stone Masonry Structures*, London: Spon.

Stell G. (2003), *Scotland's Buildings*, East Linton: Tuckwell.

Sun Z., Xie J., Zhang Y. and Cao Y. (2019), "As-Built BIM for a Fifteenth-Century Chinese Brick Structure at Various LoDs", *ISPRS International Journal of Geo-Information*, vol. 8 no. 12 p. 577.

Sutcliffe G. (1898) *Modern House Construction Vol. 1*, Glasgow: Blackie and Son.

Swailes T. (2009), *Guide For Practitioners 5: Scottish Iron Structures*, Edinburgh: Historic Scotland.

Swallow P., Dallas R., Jackson S. and Watt D. (2001), *Measurement and Recording of Historic Buildings*, Shaftesbury: Donhead.

Torney C., Schmidt A. and Graham C. (2020), *A Data Driven Approach to Understanding Historic Mortars in Scotland,* Edinburgh: Historic Environment Scotland.

Urquhart D. (2007), *Conversion of Traditional Buildings Part 1*, Edinburgh: Historic Scotland.

Vicat L.J. (1837), *Mortars and Cements*, London: Weale.

Warren J. (1999), *Conservation of Brick,* London: Butterworth-Heinemann.

Watt D. (2007), *Building Pathology: Principles and Practice*, Oxford: Blackwell.

Watt D. (2011), *Surveying Historic Buildings*, London: Routledge.

Weaver M.E. and Matero F.G. (1993), *Conserving Buildings: A Guide to Techniques and Materials*, New York: John Wiley.

Wermeil S. (1993) "The Development of Fireproof Construction in Great Britain and the United States in the Nineteenth Century", *Construction History*, vol. 9, pp. 3–26.

Wight J. (1972), *Brick Building in England from the Middle Ages to 1550*, London: Baker.

Woodforde J. (1976), *Bricks to Build a Brick House*, London: Routledge.

Young M. (2015), *Short Guide on Thermal Imaging in the Historic Environment,* Edinburgh: Historic Environment Scotland.

Index